T0260730

Public Procurement for Innovation

EU-SPRI FORUM ON SCIENCE, TECHNOLOGY AND INNOVATION POLICY

Series Editors: Susana Borrás, *Department of Business and Politics, Copenhagen Business School, Denmark,* Jakob Edler, *Manchester Institute of Innovation Research, Manchester Business School, UK,* Stefan Kuhlmann, *Science, Technology and Policy Studies, University of Twente, the Netherlands* and Ismael Rafols, *INGENIO (CSIC-UPV), Polytechnic University of Valencia, Spain and SPRU, University of Sussex, UK*

The aim of this series is to present some of the best and most original research emanating from the Eu-SPRI Forum on Science, Technology and Innovation Policy. The typical questions addressed by the books in the series will include, but not be limited to:

- What is the role of science, technology and innovation policy in the 21st century?
- How can policies cope with 'grand social challenges' in the areas of health, energy, security or the environment?
- Are there better ways to link governments' science and innovation policies to other public policies?
- What are the innovation policy rationales and instruments for successfully fostering competitiveness and economic growth?
- Which public policies help to proactively shape responsible and legitimate technological innovation?
- How can public sector research be made more creative and effective?
- How can more intelligent interactions be achieved between investments in research and higher education policies for universities?
- How can the forces of globalisation and localisation be balanced?

Titles in the series include:

The Governance of Socio-Technical Systems
Explaining Change
Edited by Susana Borrás and Jakob Edler

Public Procurement for Innovation
Edited by Charles Edquist, Nicholas S. Vonortas, Jon Mikel Zabala-Iturriagagoitia and Jakob Edler

Public Procurement for Innovation

Edited by

Charles Edquist

Professor, CIRCLE, Lund University, Sweden

Nicholas S. Vonortas

Professor, Center for International Science and Technology Policy and Department of Economics, The George Washington University, USA

Jon Mikel Zabala-Iturriagagoitia

Researcher, Deusto Business School, University of Deusto, Spain

Jakob Edler

Professor, Manchester Institute of Innovation Research, University of Manchester, UK

EU-SPRI FORUM ON
SCIENCE, TECHNOLOGY AND INNOVATION POLICY

Cheltenham, UK • Northampton, MA, USA

Published by
Edward Elgar Publishing Limited
The Lypiatts
15 Lansdown Road
Cheltenham
Glos GL50 2JA
UK

Edward Elgar Publishing, Inc.
William Pratt House
9 Dewey Court
Northampton
Massachusetts 01060
USA

Paperback edition 2016

A catalogue record for this book
is available from the British Library

Library of Congress Control Number: 2014950840

This book is available electronically in the **Elgar**online
Economics subject collection
DOI 10.4337/9781783471898

ISBN 978 1 78347 188 1 (cased)
ISBN 978 1 78347 189 8 (eBook)
ISBN 978 1 78471 362 1 (paperback)

Typeset by Servis Filmsetting Ltd, Stockport, Cheshire

Contents

List of contributors vii
Preface xv

1. Introduction 1
Charles Edquist, Nicholas S. Vonortas and Jon Mikel Zabala-Iturriagagoitia

PART I CONCEPTUAL FRAMEWORK

2. The meaning and limitations of public procurement for innovation: a supplier's experience 35
Jakob Edler, Luke Georghiou, Elvira Uyarra and Jillian Yeow
3. Building capability for public procurement of innovation 65
Ville Valovirta
4. Risk management in public procurement of innovation: a conceptualization 87
Jakob Edler, Max Rolfstam, Lena Tsipouri and Elvira Uyarra
5. Forward commitment procurement and its effect on perceived risks in PPI projects 110
Hendrik van Meerveld, Joram Nauta and Gaynor Whyles

PART II CASE STUDIES

6. Innovation and public procurement in the United States 147
Nicholas S. Vonortas
7. Public procurement for innovation elements in the Chinese new energy vehicles program 179
Yanchao Li, Luke Georghiou and John Rigby

8. Public procurement for e-government services: challenges and problems related to the implementation of a new innovative scheme in Greek local authorities 209
Yannis Caloghirou, Aimilia Protogerou and Panagiotis Panaghiotopoulos

9. Closing the loop: examining the case of the procurement of a sustainable innovation 235
Jillian Yeow, Elvira Uyarra and Sally Gee

10. Public procurement for innovation in developing countries: the case of Petrobras 263
Cássio Garcia Ribeiro and André Tosi Furtado

11. Conclusions: lessons, limitations and way forward 299
Jakob Edler, Charles Edquist, Nicholas S. Vonortas and Jon Mikel Zabala-Iturriagagoitia

Index 307

Contributors

Yannis Caloghirou is Professor of Economics of Technology and Industrial Strategy and Head of the Innovation and Entrepreneurship Unit at the National Technical University of Athens (NTUA), Greece. He has acted as a scientific coordinator in a number of European research projects in the broader area of socioeconomic research. He has served in top policy-making positions in Greece, among them as Secretary General for Industry and as Secretary for the Information Society. He has sat in a number of European Union (EU) high-level expert and policy groups, among them in the expert group on Public Procurement for Research and Innovation. He was also co-Rapporteur of the EU High-Level Policy Group on the Socio-Economic Benefits of the European Research Area. Moreover, he has extensive work experience in industry as well as in policy advisory and policy design and evaluation positions. Professor Caloghirou has written extensively on topics related to his research in scholarly journals, edited books and the popular and business press.

Jakob Edler is Executive Director of MIoIR (Manchester Institute of Innovation Research) at the Manchester Business School, University of Manchester, UK, and Professor of Innovation Policy and Strategy. His field of activity and publication comprise: evaluation and conceptual development of research, development and innovation policies, demand-based innovation policy and public procurement, European innovation policy and modes of governance; comparative research on internationalization strategies in science and technology policy; internationalization of industrial research and development (R&D) and public research, knowledge supply and technology transfer; industrial knowledge management and patent strategies. Professor Edler has advised the European Union (EU), the Organisation for Economic Co-operation and Development (OECD) and a range of governments, and published intensively on these issues. As for evaluation, he has conducted many projects for various European and regional governments and the European Commission. He has been a member of various evaluation expert groups at EU level, analysing the impact of the Framework Programme as well as individual instruments. He has also advised DG Regio on evaluation practices and lessons to be learned for innovation policy. Recently, he has led the InnoAppraisal project funded by the EU which analysed evaluation practice in national innovation policy schemes across Europe, and participated in a project on evaluation of innovation policy schemes at regional level across Europe.

Charles Edquist is Holder of the Ruben Rausing Chair in Innovation Research at CIRCLE, Lund University, Sweden, since February 2003. He is one of the founders and the first director (2004–2011) of CIRCLE (Centre for Innovation, Research and Competence in the Learning Economy) at Lund University, which by 2011 had developed into one of the largest innovation research centres in Europe, and a leading international centre for research and advice on research and development (R&D), innovation, knowledge creation, entrepreneurship and economic dynamics. Previously he held a Chair at the University of Linköping, Sweden. His comprehensive list of publications include books and articles on innovation processes, innovation systems, innovation policy and governance of innovation systems. He has a long experience in evaluations of innovation systems and innovation policy, as well as consultancy tasks with numerous international organizations including the Organisation for Economic Co-operation and Development (OECD), United Nations Industrial Development Organization (UNIDO), European Union (EU), United Nations Conference on Trade and Development (UNCTAD), International Labour Organization (ILO), United Nations Development Programme (UNDP) and United Nations Educational, Scientific and Cultural Organization (UNESCO).

André Tosi Furtado is Professor at the Department of Science and Technology Policy – Institute of Geosciences – University of Campinas (UNICAMP), in Brazil. He earned his Bachelor's, Master's and PhD in Economics at the University of Paris I. His research interests include energy and innovation policies, especially regarding Brazilian oil and ethanol industries.

Sally Gee is a Research Fellow at the Manchester Institute of Innovation Research (MIoIR) and the Sustainable Consumption

Institute (SCI), the University of Manchester, UK. Sally has broad research interests in the area of innovation studies, focusing on the dynamics of innovation over time, the emergence of new sectors and the evolution of innovation systems towards more sustainable states. She uses in-depth qualitative case studies to explore complex inter- and intra-organization interactions over time. She has published in a variety of media including books, blogs and journals such as *Technology Analysis and Strategic Management, Industry and Innovation* and the *Journal of Cleaner Production*.

Luke Georghiou BSc, PhD is Vice-President for Research and Innovation at the University of Manchester, UK, where he is responsible for the university's research strategy and performance, business engagement and commercialization activities. He holds the Chair of Science and Technology Policy and Management. His research interests include evaluation of research and development (R&D) and innovation policy (particularly in relation to the use of public procurement and other demand-side measures), foresight, national and international science policy, and management of innovation. He has chaired or been a member of several high-profile committees including the Aho Group and the European Commission (EC)'s Expert Group on ERA (European Research Area) Rationales. He has an extensive list of publications including articles in *Nature, Science* and the *Harvard Business Review*. He is a member of the Academia Europaea.

Yanchao Li is Research Associate at Manchester Institute of Innovation Research (MIoIR), University of Manchester, UK. Her research interests include demand-side innovation policies (in particular public procurement of innovation), evaluation of innovation policies, and innovation policies for catching up. She has been involved in various teaching and research activities at MIOIR including the 'Understanding Public Procurement for Innovation' (UNDERPINN) project and the 'Research Infrastructure for Research and Innovation Policy Studies' (RISIS) project. She obtained her PhD in 2013 from the University of Manchester with the thesis 'Public procurement as a demand-side innovation policy in China – an exploratory and evaluative study'. Prior to this, she obtained her Master's and Bachelor's degrees at Tsinghua University and Shanghai Jiao Tong University, China.

Joram Nauta, MSc and BEng, is a Senior Project Manager within TNO, the Netherlands. He was trained as a Civil and Industrial Engineer and worked for more than ten years on innovations in planning, design and financing of healthcare infrastructure, primarily as a Researcher and Project Manager. From 2009 onwards he has been involved in projects involving procurement of innovations, concerning forward commitment procurement as well as pre-commercial procurement.

Panagiotis Panaghiotopoulos is a PhD Candidate in economic and strategic analysis of information and communication technologies at the Laboratory of Industrial and Energy Economics of the National Technical University of Athens (LIEE-NTUA), Greece. He received his Diploma in Chemical Engineering from the NTUA and his Master's in Business Mathematics from the University of Athens and the Athens University of Economics and Business, Greece. His research interests revolve around dynamic capabilities in the public sector and particularly at a local government level, information society, public procurement for innovation, innovative entrepreneurship and environmental economics. He has contributed to related research projects financed by the European Union (EU) and Greek authorities (including the Union of Greek Municipalities and the Hellenic Federation of Enterprises).

Aimilia Protogerou holds a PhD in Business Strategy and Industrial Policy and is Research Fellow at the Laboratory of Industrial and Energy Economics of the National Technical University of Athens (LIEE-NTUA), Greece. Her research interests revolve around strategic management of technology and innovation, technology policy and cooperative research and development, innovation networks and knowledge-intensive entrepreneurship. She has contributed as a Researcher and Principal Researcher to a large number of related research projects, mainly financed by the European Union (EU) and the Greek National Secretariat of Research and Technology. She has published her work in international journals such as *Industrial and Corporate Change*, *Economics of Innovation and New Technology*, *Science and Public Policy*, *Journal of Technology Transfer* and the *European Management Journal*.

Cássio Garcia Ribeiro is Assistant Professor and Researcher at the São Paulo State University (Unesp), in Brazil. He earned his Bachelor's Diploma in Science in Economics from Unesp in 2002; his MSc in Science and Technology Policy from the University of Campinas, Brazil, in 2005; and his PhD in Science and Technology Policy from the University of Campinas in 2009. His research interests include public procurement policy, technological learning, oil and gas industry, Petrobras, and science and technology policy.

John Rigby is Senior Research Fellow at the Manchester Institute of Innovation Research (MIoIR), University of Manchester, UK. His academic interests extend around the whole policy cycle from policy and programme design and development, through implementation to evaluation, and impact assessment, but they also include methodology. John has worked extensively on policy implementation and evaluation for a range of bodies including the UK government, the European Union (DG Enterprise, DG Research and Innovation, DG Information Society and Media), most recently focusing on the implementation of the procurement of innovation. He led between 2010 and 2012 the DG Enterprise study on the feasibility of European Union (EU) support to the procurement of innovation. In the area of evaluation of science and innovation policy, John's main contributions have used bibliometric methods.

Max Rolfstam is Associate Professor at the Department of Business and Management, Aalborg University, Denmark. After finishing his PhD thesis at Lund University, Sweden, he carried out his postdoctoral research at the University of Southern Denmark. His research and teaching activities are around the field of innovation studies, with a broad interest in how institutions and institutional change affect possibilities for innovation. He has extensively worked with issues related to how public procurement can drive innovation on the European, national and regional level in a number of projects.

Lena Tsipouri is Associate Professor at the Department of Economic Sciences, University of Athens, Greece. She studied Economic Sciences at the Universities of Athens and Vienna, Austria; completed her PhD (Doctorat d' Etat), at the University of Paris II, France, which received the first prize of the year 1988; and undertook postdoctoral research with a Fulbright Fellowship at MIT, Cambridge, MA, USA. Professor Tsipouri teaches economic development, European economic integration, the economics of technological change and theory of the firm. Her scientific research and publications are in the topics of research and innovation, regional development and corporate governance. Her most recent publications include journal articles on the public procurement of innovation, comparing policies of science in society and innovation, articles on the Greek economic crisis and issues on diversity. Professor Tsipouri is professionally involved in consulting work for European Union (EU) institutions, the Organisation for Economic Co-operation and Development (OECD) and the United Nations (UN) on the same topics as her scientific research, and is a member of several professional associations. She is currently the chairperson for the EU Innovation for Growth Group.

Elvira Uyarra is Senior Lecturer at Manchester Business School, and a member of the Manchester Institute of Innovation Research, UK. Her research and teaching experience sit broadly within in the area of innovation studies, policy studies and economic geography. Her research interests include the conceptualization of innovation systems (particularly regional innovation systems), various dimensions of innovation policy (policy rationales, multilevel governance, policy dynamics and the idea of the policy mix), and demand-side innovation policies (particularly public procurement). Her work has been published in leading journals in geography and innovation studies such as *Research Policy, European Planning Studies, Research Evaluation, Journal of Cleaner Production, Environment and Planning C* and *Research Evaluation*, the *Journal of Evolutionary Economics*, as well as in edited volumes published by Palgrave, Edward Elgar and Elsevier.

Ville Valovirta is Senior Scientist at VTT, Technical Research Centre of Finland. He holds an MS (Pol.) from the University of Helsinki, Finland. His research focuses on innovation systems, technology foresight, innovation policy and public–private collaboration for innovation. He is pursuing research on diffusion of urban innovation in cities and public procurement of innovation. Valovirta has carried out foresight projects in Finland, Australia, South Korea and Chile. He has conducted several programme evaluations and impact assessment related to innovation policy, research and development (R&D) programmes and commercialization support schemes. Valovirta has a background in a private consultancy where he held responsibility for management of policy analysis and evaluation services.

Hendrik van Meerveld, MSc has a degree in civil engineering and management with a specialization in construction process management. Both as a Researcher and a Consultant, he has been involved in several projects relating to procurement and innovation (including forward commitment procurement and pre-commercial procurement).

Nicholas S. Vonortas is Professor of Economics and International Affairs at the George Washington University in Washington, DC, USA. He is a faculty member of the Department of Economics and of the Center for International Science and Technology Policy (CISTP). Professor Vonortas's teaching and research interests are in industrial organization, in the economics of technological change, and in science, technology and innovation policy. He specializes on strategic partnerships and innovation networks, theory of investment under uncertainty, technology transfer, knowledge-intensive entrepreneurship and high-risk financing, and the appraisal of the economic returns of research and development (R&D) programmes. Professor Vonortas is the co-editor of the peer-reviewed journal *Science and Public Policy*. He holds a PhD and MPhil in Economics from New York University, USA; an MA in Economic Development from Leicester University, UK; and a BA in Economics from the University of Athens, Greece.

Gaynor Whyles is Director of JERA Consulting, a UK-based consultancy specializing in innovation, and is passionate about the role of innovation procurement in improving public services and creating economic opportunities. JERA provides consultancy and training in public procurement for innovation (PPI) and innovation to the public and private sector. Gaynor has worked as an Independent Consultant for the UK Department for Business, Innovation and Skills (BIS) in the field of innovation procurement since 2005. She was responsible for developing forward commitment procurement (FCP) concept into a practical tool for innovation procurement and has initiated and managed successful FCP demonstration projects. She sits on the DG Enterprise PPI Platform Advisory Group. A graduate of Imperial College, University of London, UK, she previously worked for the UK Research Councils in the UK and Brussels, Belgium.

Jillian Yeow is a Research Associate at the Manchester Institute of Innovation Research (MIoIR) at the University of Manchester, UK. Her research interests are broadly in the area of organizational innovation and innovation management, particularly in impacts of technologies on work and organizational practices. Her work mainly uses qualitative case studies to explore tensions, behaviours and experiences. Jillian has published in books and journals such as the *Journal of Public Procurement* and *Technological Forecasting and Social Change*.

Jon Mikel Zabala-Iturriagagoitia is a Researcher at the University of Deusto in San Sebastián, Spain and Assistant Professor at CIRCLE, Lund University, Sweden. His research and teaching interests are related to the fields of innovation policy and innovation management. As a Researcher he has contributed to the development of methodological approaches for the assessment of innovation potential, innovation policy instruments such as public procurement for innovation and pre-commercial procurement, and innovation management tools (in particular, creativity and technology watch). As a Lecturer, he has been involved in courses at the PhD, MSc and undergraduate levels at several European and Latin American universities. His work has been published in journals such as *European Planning Studies, Regional Studies, Scientometrics, Research Policy, Science and Public Policy, R&D Management* and *Technovation*.

Preface

Every book has its own history of authors, structure and contents. The story of this book began back in January 2009, when three of the editors (Edquist, Vonortas and Zabala-Iturriagagoitia) met in Milan, Italy at the kick-off meeting of the AEGIS project (Advancing Knowledge-Intensive Entrepreneurship and Innovation for Economic Growth and Social Well-being in Europe). It was a 7th Framework Programme in which the three were involved as project partners.

One of the work packages of that project was called 'The Relations between Demand and Public Innovative Procurement and between Knowledge Intensive Entrepreneurship and Innovation'. As a result of this work, Edquist, Vonortas and Zabala-Iturriagagoitia developed a robust relationship and a common understanding of the role that public procurement could play as a demand-side instrument in relation to innovation. When the project was about to end, the three researchers met again in one of the final AEGIS meetings and discussed ways and means of developing this research avenue. The idea of writing a joint book on public procurement for innovation popped up, and 'the book project' was born.

The goal was to reach out to a wider set of scholars who could contribute to such a book. Edquist and Zabala-Iturriagagoitia then organized a Special Session dealing with Public Procurement for Innovation at the 2012 Annual Conference organized by the EU-SPRI (European Forum for Studies of Policies for Research and Innovation) network, which took place in Karlsruhe (June 2012). The authors included in this book contributed papers to that Special Session.

The group interested in contributing to the book then had discussions addressing such issues as the title, contents, possible weaknesses, and so on. When the work entered a more intense stage, Vonortas was invited to spend a couple of months at CIRCLE (Centre for Innovation, Research and Competence in the Learning Economy) in Lund in early spring 2013, and the three of us (Edquist, Vonortas and Zabala-Iturriagagoitia) wrote the Introduction to the book. Jakob Edler, who had had a leading role in the discussions since Karlsruhe, was also invited to participate as an editor in late spring 2013. He very kindly accepted the invitation. Also Jaya Reddy has greatly contributed to the clarity of the result, thanks to his precise improvement of our English.

This little history helps to explain why only three of the editors are credited with the Introduction, while the concluding chapter is attributed to all four. It also explains the order in which the editors of the book are listed.

Several reviews of the submitted manuscripts were undertaken between June 2012 and December 2013, leading to substantial work by the editors and the respective authors. Now, all the chapters have been reviewed, all the contents have been edited and we are in a position to write these few lines that hopefully provide the reader with a clearer view of the journey we embarked upon back in 2009.

Charles Edquist, Nicholas S. Vonortas, Jon Mikel Zabala-Iturriagagoitia and Jakob Edler
Lund, Sweden; Washington, DC, USA; Donostia-San Sebastian, Spain; Manchester, UK
April 2014

1. Introduction

Charles Edquist, Nicholas S. Vonortas and Jon Mikel Zabala-Iturriagagoitia

INTRODUCTION

This book addresses public procurement for innovation, which is a demand-side innovation policy instrument in the form of an order, placed by a public organization, for a new or improved product to fulfill its particular needs. With this book we aim to contribute to the understanding of public procurement for innovation, including its operationalization and implementation. It is oriented to a wider readership of scholars interested in the study of innovation policies, practitioners involved in the public procurement of goods and services, and students of public policy.

Traditionally, innovation policy initiatives have mostly come from the supply side. Countries and regions have actively implemented and used innovation policy instruments such as fiscal measures, support for training and mobility, public financing of research and development (R&D), information and brokerage support and networking measures.

The role of demand as an enabler and source of innovation has been a topic in innovation studies and innovation policy for quite some time (Izsak and Edler, 2011). Discussions on the positive impacts of demand-side innovation policies took place as far back as the 1970s (Dalpé, 1994; Geroski, 1990; Mowery and Rosenberg, 1979; Rothwell and Zegveld, 1981). Recent years have seen a resurgence of the interest in demand-side approaches to innovation policy (Edquist and Hommen, 1999; OECD, 2011). For example, in 2004, three governments issued a position paper to the European Council calling for the use of public procurement across Europe to spur innovation (Edler and Georghiou, 2007; French, German, UK Governments, 2004). This development continued and was manifested in various reports, including the Aho Group Report (Aho et al., 2006), which identified several application areas, or grand challenges, where demand-side policies could be used to a larger extent: e-health, pharmaceuticals, energy, environment, transport and logistics, security and digital content (Edler and Georghiou, 2007, p. 951).

Demand-side intervention is intended to increase the demand for innovations, to improve the conditions for the uptake of innovations and to improve the articulation of demand (Edler, 2007). Cluster policies, regulation (for example, standards), public procurement (that is, R&D procurement and innovation procurement), and support of private demand are examples of demand-side innovation policy instruments (Edler and Georghiou, 2007, p. 953). We largely limit the scope in this book to the analysis of public procurement as an innovation policy tool, that is, what we refer to as public procurement for innovation (PPI).

Until fairly recently this phenomenon was called 'public technology procurement'. Edquist et al. (2000) compiled an extensive set of cases illustrating the potential for innovation involved in this policy instrument. Since then, the terminology of the 1990s and earlier has changed into 'public procurement for innovation'. This reflects a widening of the content of the notion, since 'innovation' includes more than 'technology'.

The next section briefly addresses the development of innovation research and innovation policy of relevance for our topic. It includes a discussion of innovation systems, determinants of innovation processes and demand-side innovation policies. The third section is devoted to presenting a taxonomy of different kinds of innovation procurement. Then we look, in some detail, at the challenges involved in the implementation of innovation procurement, such as identification of challenges to be mitigated by public procurement for innovation, translation of these into functional requirements, and the tendering process (including the assessment of the tenders and the delivery process). The final section concludes the Introduction with a short summary of the chapters in this book.

DETERMINANTS OF INNOVATION AND DEMAND-SIDE POLICY INSTRUMENTS

This section briefly considers the determinants of innovation processes and demand-side innovation policy instruments. However, first let us define what we understand by innovation.

Innovations are new creations of economic or societal significance mainly carried out by firms. Innovations are the result of interactive processes among the multiple actors contained within innovation systems (Edquist, 1997). In fact, according to the innovation systems approach, interaction and interactive learning between organizations are pillars upon which innovations are based. These innovations may be new or improved products or processes. While new or improved products may be material goods or intangible services, new or improved processes may be technological or organizational advancements.

To qualify as an innovation, the product or process needs to be implemented (Oslo Manual, 2005, p. 47), which means that the new products must be introduced on a market or that the new processes are used in production. Innovations must therefore be commercialized, that is, launched on the market, or in other ways widely diffused to users on a large scale in the economy or society. This includes innovations in public activities. A prototype is a model that functions under certain conditions. It is not produced in a large series, not commercialized and its commercial viability is not proven. Hence, a tested prototype is not an innovation.

In the early days of innovation research and innovation policy, the so-called 'linear model', based on the assumption that innovations are applied scientific knowledge, dominated the view of how innovation processes developed. It was called 'linear' because the process was seen as a number of well-defined and consecutive stages that innovations were assumed to go through, for example basic research, applied research, development work resulting in new products and processes that ultimately influence growth and employment. It was a supply-push view.

In the realm of innovation research, the linear view has been replaced almost completely by the systems of innovation approach in the latest couple of decades. In its different versions, it is defined in terms of determinants of innovation processes, although different determinants are emphasized in different versions. A general definition of systems of innovation is that they include 'all important economic, social, political, organizational, institutional and other factors that influence the development, diffusion and use of innovations' (Edquist, 1997, p. 14).[1]

One way to introduce the variety of innovation policy instruments is to talk about the activities that configure an innovation system, which are the same as the determinants of innovation processes. Table 1A.1 in the Appendix contains a list of important activities in innovation systems. The activities are not ranked according to importance, but are clustered as:

- provision of knowledge inputs to the innovation process (for example, research);
- demand-side activities (for example, public procurement for innovation);
- provision of constituents in innovation systems (for example, entrepreneurship);
- support services for innovating firms (for example, public seed funding of innovations).

The list of activities presented in Table 1A.1 (also sometimes called functions[2]) is preliminary, hypothetical and one among several (Johnson, 2001; Edquist, 2005; Hekkert et al., 2007). It will certainly be revised as our knowledge about the determinants of innovation processes increases. It is important to point out though that public innovation policy is not included as one of these ten activities. The reason is simply that public policy is a part of all ten activities. Part of each activity is performed by public organizations, which is policy (see the definition of innovation policy below). What is important is the division of labour between private and public organizations with regard to the performance of each of the activities.

Non-firm public organizations do not normally take part directly in innovation processes, although they certainly are important in terms of participating, for example, in research and other activities that influence innovation (see Table 1A.1). They affect the context in which the innovating firms operate. Innovation policy may thus be understood as actions by public organizations that influence innovation processes, that is, the development and diffusion of innovations (Edquist, 2011, p. 1728)

What then is this context? A general answer to this question is that the context is all those things that influence innovation processes; that is, all the determinants of innovation processes. The literature on systems of innovation has discussed the determinants of innovation processes extensively (Galli and Teubal, 1997; McKelvey, 1997; Edquist, 1997; Johnson, 2001; Edquist, 2005; Hekkert et al., 2007; Bergek et al., 2008). Innovation policy can thus be understood as the actions by public organizations that influence innovation processes.[3] Such actions relate to all key activities in systems of innovation, see Appendix Table 1A.1 (Edquist, 2011).

In the realm of innovation policy, the linear view is still much more dominant than it is in innovation research (Godin, 2006). In recent years there has been an increasing interest in 'broad-based innovation policies', systemic innovation policies, 'a demand-pull view', and 'demand-oriented policy instruments', such as public procurement for innovation (PPI) or pre-commercial procurement (PCP) (Edquist and Zabala-Iturriagagoitia, 2012, forthcoming). This may constitute the very beginning of a transformation towards a 'holistic innovation policy'. This is defined as 'a policy that integrates all public actions that influence or may influence innovation processes', for example by addressing all the ten activities in Table 1A.1 in a coordinated manner. It must include demand-side innovation policy instruments (Edquist, 2014b).

Edquist (2014c) shows that most European Union (EU) member countries are trying to develop a holistic innovation policy, but that very few have come close to achieving this. A questionnaire was sent to 23 EU member states in 2014; 19 responded. That the linear model is still dominant in innovation policy was strongly confirmed by the responses. Sixteen of the 19 countries (84 per cent) indicate that they are striving in the direction of developing innovation policy into a more holistic one. However, the responses also indicate that they are not actually pursuing much innovation policy that can be considered demand-side-oriented. Also, a majority of the countries indicate that 'provision of R&D results' is the most important activity in terms of resources spent for innovation policy purposes. Together, these responses clearly indicate that many of the countries striving in the direction of pursuing a holistic innovation policy have a long way to go on the path from linear to holistic. There may also be countries that are paying lip service to holism, but have actually not achieved much at all (Edquist, 2014c).[4]

Demand-side innovation policies can be defined as 'a set of public measures to increase the demand for innovations, to improve the conditions for the uptake of innovations or to improve the articulation of demand in order to spur innovations and the diffusion of innovations' (Edler and Georghiou, 2007, p. 952).

Table 1A.2 in the Appendix provides a list of instruments that can aid the development of demand-based innovation interventions. These are grouped into three major blocks. The first one addresses the purchasing power of the public sector, an area where public procurement plays a major role. The second block deals with instruments that support demand, but on the private side. Three other types of instruments are included in this block: direct support, indirect support and regulations. Finally, the third block entails all the possible combinations of both supply- and demand-side instruments.

We do not intend to address the whole battery of all demand-side innovation policy instruments in this book. Instead, we limit ourselves to innovation procurement, mainly PPI.

In the real world, however, the instruments of innovation policy are rarely used standing 'alone'. Normally innovation policy instruments are combined in specific mixes, using groups of different instruments in a complementary manner. Instrument mixes are created because the solution of specific problems requires complementary approaches to the multi-dimensional aspects of innovation-related problems (Borrás and Edquist, 2013; Flanagan et al., 2011; Guy et al., 2009; Nauwelaers, 2009).

PUBLIC PROCUREMENT FOR INNOVATION: A TAXONOMY[5]

There are different types of public procurement (for innovation). The different types are used as policy instruments in different ways and according to different rules and procedures. They are also often mixed up with each other.

Public procurement refers to the purchase of goods and services by a public agency. In the case of regular procurement, public agencies buy ready-made products 'off the shelf', and normally no innovations are the result of the public intervention. Public procurement is roughly estimated to account for 15–20 per cent of gross domestic product (GDP) in developed countries. In 2009 it was 19.4 per cent of European GDP, equal to €2.3 trillion.

Public procurement for innovation (PPI) occurs when a public organization places an order for the fulfillment of certain functions (that are not met at the moment of the order or call) within a reasonable period of time through a new or improved product.[6] Hence, the objective of PPI is not primarily to enhance the development of new products, but to target functions that satisfy human needs, solve societal problems or support agency missions or needs. Still, some form of innovation (new product or process) is necessary before delivery can take place.

Because of the large numbers involved, and to the extent that something new is purchased, there is also the notion that public procurement may provide a 'lead customer' or a lead market for an innovative product or process (European Commission, 2007a).

We must point out here that the diffusion of the product from the procuring organizations is not always among the major objectives of this type of programme. However, there are cases in which diffusion of the new product is aimed at from the very start of the procurement process. This difference reflects the distinction between PPI carried out mainly for the missions or needs of the procuring agency, and PPI to support economy-wide innovation. That way, a public agency may demand the purchase of certain products or systems that are novel to the agency, but not to the market. Be that as it may, innovation is needed in all PPI before delivery can take place.

PPI is a part of the second set of key activities in innovation systems in Table 1A.1, namely demand-side activities. It provides a demand-pull that complements the supply-push for innovation, which has traditionally tended to attract attention in innovation policy studies, and is an important demand-side innovation policy instrument (Dalpé, 1994; Edler and Georghiou, 2007; Edquist et al., 2000; Geroski, 1990; Rothwell and Zegveld, 1981).

A taxonomy that covers different types of PPI is presented below. We include the notion of pre-commercial procurement (PCP) although it is not PPI.[7] First, however, we present the notion of innovation-friendly public procurement, although it is not PPI either.

Innovation-friendly public procurement is regular procurement which is carried out in such a way that new and innovative solutions are not excluded or treated unfairly. The background is a concern that many public procurements are carried out in a routine-like manner, meaning that the procuring organization demands the same solution as in the previous procurement. This might actually constitute an obstacle to innovation. The most powerful means to overcome this obstacle is to require calls for public procurement to be formulated in functional terms, and not as descriptions of products.[8] Hence, there is reason to distinguish between innovation-friendly procurement and public procurement for innovation. Innovation-friendly procurement does not require innovation as PPI, but encourages and facilitates innovation (see the section below on 'Operationalizing public procurement for innovation', and Edquist, 2014a).

The classification of PPI below is made according to three dimensions. The first dimension that allows us to identify different forms of PPI relates to whom the user of the result (good, service, system, and so on) might be. This dimension may be used to identify two categories of PPI: direct and catalytic (Edquist et al., 2000; Hommen and Rolfstam, 2009; Edquist and Zabala-Iturriagagoitia, 2012).

Direct PPI is when the procuring organization is also the end-user of the product resulting from the procurement, which is the 'classic' case. The buying agency simply uses its own demand or need to influence or induce innovation. This type of PPI includes the procurement undertaken to meet the ('mission') needs of the public agencies themselves. It exerts direct 'demand-pull' on suppliers, often through long-term contracting arrangements. Nonetheless, the resulting product is often also diffused to other users once the initial procurement process is finished, and the agency has benefited from the results obtained for a certain period. Hence, innovations resulting from PPI can be useful for the procuring agencies, as well as for society as a whole.

Catalytic PPI is when public sector organizations act as buyers even if they are not the intended end-users of the results of the procurement process. In other words, the procuring agency serves as a catalyst, coordinator and technical resource for the benefit of end-users. The needs are located 'outside' the public agency acting as the 'buyer'. Thus, the public agency aims to procure new products on behalf of other organizations, and public demand articulates, sponsors, and helps to shape private demand. It acts to catalyze the development of innovations for broader public use and not for directly supporting the mission of the agency.

The second dimension in the classification relates to the character of the innovation embedded in the resulting product. This dimension leads to two types of procurement: incremental and radical (Edquist et al., 2000).

Incremental PPI is when the product or system procured is

adaptive and new only to the user of the results of the procurement process (public agency, private firm, country, region, city, and so on). Hence, innovation is required in order to adapt the product to specific national or local conditions. It may also be labelled 'diffusion-oriented' or 'absorption-oriented' PPI.

Radical PPI implies that completely new-to-the-world products and/or systems are created as a result of the procurement process. It may be regarded as 'creation-oriented' PPI and involves the development of brand new innovations.

It is also important to address a third dimension in the classification, namely the fact that PPI can be characterized by different degrees of collaboration and interactive learning (among procurers, suppliers and – sometimes – other organizations). This collaboration is a matter of degrees, not a dichotomous variable. It is important, since we know that interactive learning is a central determinant of the development and diffusion of innovations. All four previous categories can be carried out with different degrees of collaboration. It may be expected, though, that collaboration is more important in catalytic than in direct PPI, simply because the catalytic type involves more than two actors.

In January 2014, the European Parliament decided on new directives on public procurement. In addition to considering the lowest price in the procurement, other dimensions are now important in the selection of contractors: quality, sustainability, social conditions and innovation. The decision also includes a new 'procedure' called innovation partnerships. Such partnerships make possible collaboration between the procuring organization and suppliers in order to achieve the objectives of the procurer. The new directive has to be implemented in all member states within 24 months.

Pre-commercial procurement (PCP) refers to the procurement of (expected) research results and is a matter of direct public R&D investments, but not actual product development. Moreover, it does not involve the purchase of a large number of units of a (non-existing) product, and no buyer of such a product is therefore involved in the procurement (as in PPI). It is not PPI since a product must be commercialized to constitute an innovation.[9] This type of procurement may also be labelled 'contract' research, and may include the development of a product prototype.[10] This type of public R&D funding is very problem-oriented and targeted, as opposed to general public R&D funding or tax deductions that firms can make for their

R&D expenditures. Of course, the procured research results may be developed into a product innovation when the PCP process (or phase) has been completed. In addition, a PCP may be the same as the R&D phase in a PPI.

OPERATIONALIZING PUBLIC PROCUREMENT FOR INNOVATION

There are many challenges involved in the implementation of PPI, and we address a number of these in this section. The typical PPI process may be divided into the following stages (Edler et al., 2005; Edquist and Zabala-Iturriagagoitia, 2012; Expert Group Report, 2005):

1. Identification of a public agency mission need or of a grand challenge and its formulation in terms of a lack of satisfaction of a human need or an unsolved societal problem.
2. Translation of the identified challenge into functional specifications.
3. Tendering process:
   a. opening of the bidding process through a tender;
   b. translation of the functional specification into technical specifications by potential suppliers;
   c. submission of formal bids by potential suppliers.
4. Assessment of tenders and awarding of contracts.
5. Delivery process:
   a. product development;
   b. production of the product;
   c. final delivery to the purchasing agency.

This general structure does not imply by any means that the PPI process is of a linear nature. As we will see in the cases illustrated in this book, these general steps are very much inter-related and intertwined. But let us see what is involved in each stage.

Identification of (Social or Agency) Challenges or Needs

There are several issues involved in (societal) needs anticipation. Who is actually responsible for performing that function? And how? Is it done through some formal, open, participatory process – through technology foresight, road mapping, public consultation, and so forth – or through more closed processes? How far into the future should one look? More concretely, is the public agency looking within a single technology life cycle (incremental PPI) or across two different life cycles (that is, creating a new one: radical PPI)? That is to say, the term 'reasonable' should be interpreted as relative to the degree of newness or radicality of the pursued solution, as is done in our understanding of PPI. It is fairly uncontroversial that public sector involvement may be warranted in the case of needs relating to 'big issues' (for example, grand challenges) where the costs are large and the risks considerable. There are several reasons for this, including:

- 'big issues' impact upon large numbers of constituents or even the whole society;
- the public sector has means of raising large amounts of funds, unavailable to the private sector, to target long-term issues;
- the public sector has means of sharing the risk among much larger numbers of bodies than private sector organizations, indeed among the whole society (all taxpayers);
- the public sector can deal much better than the private sector with the adverse effects of widespread positive externalities on the incentives to innovate and the consequent free-riding phenomena;[11]
- the private sector tends to be much better and faster at reading market signals for 'smaller' (near-term) innovations which entail limited and measurable risk, in marked contrast to 'bigger' (long-term) innovations which entail unmeasurable uncertainties.[12]

The ideal outcome of the exercise to identify (social) challenges and needs is the creation of a public 'vision', which has as an important by-product the transformation of genuine uncertainty confronting the private sector into manageable risk. Governments, politicians, administrations, policy-makers and public agencies may specify long-term, or at least mid-term, objectives in sectors including energy, transport, health, communication and defence (all suffering from large externalities and uncertainties). In this way future public demand may be influenced and widely communicated.[13] While it is extremely hard to predict demand in the longer term, this may help to transform uncertainty to limited risk in relation to public demand. The state may also adapt its own R&D funding to accommodate these objectives.[14] This could induce private suppliers to respond by investing in R&D and product development in the direction indicated by the probable future demand.

A case in point here is the need for interactive learning between organizations as an extremely important factor for innovations to emerge, an issue emphasized in the innovation systems literature (see the section on 'Determinants of innovation and demand-side policy instruments'). One way to achieve such interaction is to organize focus groups or networks of stakeholders to help articulate the social need and formulate the 'vision' within certain need or problem procurement areas. They should involve potential users, politicians, policy-makers (high-level bureaucrats), researchers, firm representatives, and so on. The researchers should come not only from the relevant fields of natural science and technology, but also from the social sciences such as economics, psychology and political science. The firm representatives should come from different divisions of firms including R&D, marketing, management, and so on. Diversity is key in order to achieve the 'new combinations' that underlie innovation in Schumpeter's terminology.[15]

Translation of the Identified Challenges into Functional Requirements

The successful 'translation' of needs or problems into functional requirements presupposes highly developed competences on the part of the procuring organization (Georghiou et al., 2013). The functional specifications must constitute solutions to the challenges while also being achievable given the state of the art at the time. Some relevant questions in this context are (Edquist, 2009):

1. For which sectors of production are the identified needs or problems relevant?
2. For which kinds of firms (large, small, old, new) are they relevant?
3. What 'visionary' products, produced by the firms in these sectors, can contribute to satisfying the needs and solving the problems?

The issue at hand is not only a matter of articulating demand and transforming it into functional requirements. This articulation must also be matched with supply possibilities. The satisfaction of the demand and the solving of the current and expected future problems must be within reach, within the specific time and budget limitations. Utilizing the system to do this smartly is often a very difficult task requiring the systematic training of PPI administrators.[16] There are ways in which this matching can take place but the process remains more of an art than a science. The focus groups or networks of stakeholders mentioned in the previous subsection may also be quite useful in this translation of needs and problems into functional specifications.

Long experience has shown that the procuring agency should limit itself to clearly articulating the need it is trying to fulfill while letting the bidders propose the best ways to fulfill this need. In other words, the procurer should abstain from providing technical specifications.[17] PPI should be pursued as 'functional procurement'. Accordingly, the procuring organization should formulate its requirements in functional terms and not in product terms. In other words, the function to be achieved should be defined, instead of defining the product to achieve it. This is a way to develop the creativity and innovativeness of the potential supplier (Edquist, 2014a).

But how is that done in practice and by whom? Ample experience in both the private and public sectors points to the usefulness of engaging prospective bidders early on, and consulting with them on a regular basis regarding what is feasible within a set time limit and budget, and what is not.[18] Expert advisory bodies can also provide great assistance here, perhaps working in parallel with public consultation.

Important threats here originate in 'special interests'. On the one hand, these may be the interests of politically influential organizations or individuals who are heavily involved in particular existing technological solutions (trajectories) and, as a result, might try to steer the process conservatively in their favour. On the other hand, there may be special interests of other societal groups that feel constrained by the current technological trajectories and are looking for new and creative solutions. Public authorities must balance such interests across the various stakeholders. But this is easier said than done. The danger of corruption is severe.

Tendering Process

Here we reach certain areas where most public agencies face significant bottlenecks. As indicated earlier, this tendering process is subdivided into three related phases.

The first phase is opening of the bidding process through a tender. This activity requires skills and is checked by demands for equality and fairness of treatment. Missteps entail serious dangers of legal challenge (lawsuits) which can be costly and consume valuable time.

The tendering process and the different phases included in it are very much dependent on the regulations that may exist in the different countries in relation to public procurement processes. In the European case, the current rules that apply to public procurement are included in Directive 2004/17 on procurement in water, energy, transport and postal services sectors, and Directive 2004/18 on procurement contracts for public works, public supply and public service. However, a set of new legislative rules was adopted by the European Parliament in January 2014. This new directive aims at replacing the current public procurement directives just mentioned.

Some of the proposed changes include increasing the flexibility and simplification of the procedures, the possibility of using life-cycle costing as an assessment criterion, the clarification of when cooperation between public bodies is subject to public procurement rules, and the possibility of conducting market consultations prior to the launch of the formal procurement procedure (COM/2011/896; European Commission, 2011). These new rules, for example, 'innovation partnerships', are also briefly mentioned in the section on 'Public procurement for innovation: a taxonomy'.[19]

In the case of the US, the Federal Acquisition Regulations of the General Service Administration are studied in detail in Chapter 6, devoted to the analysis of PPI in the United States. Chapter 7 discusses the most relevant rules that apply to public procurement intervention in China.

The second phase is translation of the functional specification into technical specifications. There are three (related) issues here: the extent of the translation to be carried out by the public sector employees; the problem of PPI risks and employee incentives pointing in different directions; and the tolerance of failure by the policy decision-making system.

As stressed above, if the objective of public procurement is to foster innovation as a means to target social and/or agency needs, the public buyer must avoid the translation of desired functionalities into technical specifications. This translation must be done by the potential supplier. Regardless, such practice is antithetical to the incentives of public purchasing managers who are inclined to minimize risk exposure. Risk reduction can be achieved either by procuring off-the-shelf products or by determining technical specifications in detail.

By definition, in PPI schemes the procurement of existing products should be partly replaced by the procurement of results in terms of societal problem-solving and the satisfaction of needs. In the short run this might be incompatible with the objectives of various agencies and budget constraints, but in the longer run it might lead to large cost savings. There is a clear tension here, and as some of the case studies in this book show, more recent trends in public procurement practice may be moving against PPI. In order to incentivize public employees (bureaucrats, policy-makers) to take on more risk, they must be protected by publicly elected officials (politicians). It is the latter individuals who should assume the risks since taxpayers' money is involved and they express the public interest.

Coming to the tolerance of failure by the policy decision-making system, we have a very interesting phenomenon. On the one hand, the emergence of new analytical methodologies nowadays allows the construction of complex project portfolios which spread risk and can thus allow the pursuit of a few highly risky projects among them (which presumably could correspond to PPI targets). On the other hand, in pluralist political systems, public media will typically pick on single failures and hammer elected officials on that basis rather than on the existence of the balanced portfolio of projects. It requires excessive political skill by an elected official to redirect attention to the whole portfolio.

The third phase, the submission of formal bids by potential suppliers, may be the area of least danger for public authorities. This stage is usually regulated by strict procedures which ensure proper handling of the submitted proposals.

The new EU procurement directives stress that supporting the development of innovation and guaranteeing the participation of small and medium-sized enterprises should be explicitly targeted by procurement initiatives. Still, there are no separate rules for regular procurement and PPI and the proposal for a new

EU directive on public procurement does not solve this problem either (Edquist and Zabala-Iturriagagoitia, 2012; Edquist, 2014a).

Assessment of Tenders and Awarding of Contracts

This is another challenging area for public authorities. The first challenge is to construct the appropriate committees with members knowledgeable enough to make intelligent decisions on the proposals. At least two kinds of competencies are required for this assessment task: technical judgement and economic judgement. However, these do not usually go hand in hand with each other, as they typically do not reside within the same expert.

The second challenge is the appropriate contracting instrument(s) for PPI. Experience shows that the special features of technology and innovation – especially relating to appropriability and uncertainty – require special treatment in terms of contracting instruments that create the appropriate incentives for private organizations to engage without manipulating the system. Here the experience of the military procurement by developed countries may prove quite useful as defence acquisitions have traditionally concerned technologically advanced, systemic products.

The third challenge has to do with the question of multiple sources versus one contractor as per the stipulations of the EU rules. Multiple sourcing in the earlier stages of longer-term innovative efforts has proven to be an effective strategy in the private sector in the presence of uncertainty.[20] In fact, in December 2007, the European Commission issued a new communication in order to introduce a new policy instrument, labelled 'Pre-commercial procurement' (European Commission, 2007b). The goal of pre-commercial procurement is to procure R&D services. Its formulation may be regarded as an exemption to current procurement regulations in Europe, as the previously discussed public procurement directives do not apply to pre-commercial procurement (see Art. 16f of 2004/18/EC, Art. 24e of 2004/17/EC). This particular form of procurement increases the scope for multiple sourcing of R&D results, and therefore partially limits this third challenge. However, pre-commercial procurement will not be addressed in detail in this book.

The last challenge is related to contract follow-up, monitoring and evaluation (Edler et al., 2012). Monitoring experts for PPI contracts require a deep understanding of the process of R&D

and technology solution development in specific technical fields. 'Discrete' monitoring should check progress regularly without becoming intrusive in the research process and experimentation. Again, there may be lessons to be learned from successful military procurement.

Delivery Process

This final stage encompasses product development, production and final delivery to the purchasing agency. The big problems here are, of course, cost overruns and time overruns. These problems affect public procurement in general but they become much more acute in the case of PPI due to the uncertainties of the targeted programmes. In some sense, there is a very strong positive relationship between the complexity of the procured product and the risk of cost and time overruns.

One huge concern here is the interference of the political system which may dictate changes in the original design to accommodate 'fluid' political needs.[21] While reacting on impulse would lead one to prescribe no such interference, the reality is that such interference is almost unavoidable in most big public projects, and what is more difficult is that it is even necessary to some extent, given the evolving needs of society.

A second concern here is technological uncertainty. If we are looking for radical, game-changing technologies or products, we run a significant risk that our early understandings were simply wrong, leading to cost and time overruns, and occasionally to outright failures. Lines of defence exist – good management systems, independent expert committees, citizen voting, and so on – but they are not invulnerable against strong external interests.

Finally, another more subtle issue of political economy ought to be pointed out with regards to the practical implications of the systematic introduction of demand-based innovation policies. Incorporating PPI on a large scale may well intensify governance difficulties and the risks that policies may push in different directions and become contradictory rather than coordinated (IDEA, 2013). Innovation policies have traditionally been (and are still[22]) quite linear and supply-side oriented. Confronted with the relative novelty (from an institutional and organizational viewpoint) of demand-side policies, bureaucracies, which are accustomed to dealing with supply-side-oriented innovation policies, may have learning difficulties. This may lead to governance difficulties, at least in a start-up phase.

Our discussion in this section has until now focused on the challenges and constraints associated with PPI practice and the behavioural and incentive structures in the public sector that accommodate them. However, a major barrier to PPI implementation could also prove to be the much higher dispersion of stakeholders and organizations. While supply-side R&D and innovation policy development and implementation are usually concentrated in only a few public agencies, this is not the case with a demand-side policy instrument such as PPI. The actual procurement projects must be handled by a large number of public agencies. Consequently, substantial learning efforts are required within the public sector at large to cope with PPI as such, and the increased coordination challenges (Navarro and Magro, 2013).

Turning to the risks of policy misalignment, the integration in the policy instrument mix of a set of new measures or instruments geared towards new objectives will require the redesign of the policy system (Borrás and Edquist, 2013). In addition to supply-side policy instruments, demand-side measures must be compatible with instruments and objectives pertaining to competition policy, social policy, fiscal policy, trade policy, sectoral policies, and so forth. The integration of an important and diverse new element into the system will change the system itself, modify interrelations within the system, cause problems of incompatibility, and intensify the risks of trade-offs. This again requires lengthy adjustments and policy learning from the side of the public sector (IDEA, 2013).

SUMMARIES OF BOOK CHAPTERS

The chapters in this book reflect on the issues discussed in this Introduction, and illustrate different types of procurement interventions. The book thus intends to provide both case studies and conceptual contributions that help to extend the frontier of our understanding in areas where there are still significant gaps. In this sense, Part I of the book (Chapters 2–5) has a more conceptual or theoretical orientation, whereas Part II (Chapters 6–10) reflects on specific cases related to PPI practice.

Chapter 2, by Jakob Edler, Luke Georghiou, Elvira Uyarra

and Jillian Yeow, discusses the meaning and limitations of PPI from the point of view of UK suppliers. On the basis of data from 800 responding suppliers in the UK, they show that the public sector is indeed a driver and a source of innovation for suppliers, especially with regard to service innovation. The study shows the range of barriers to innovation, such as the lack of application of those procurement practices that are most conducive to innovation, the general attitudes towards risk and innovation, and procurers' lack of market and technological knowledge. It also highlights some important differences between types of suppliers, sectors and government areas, and concludes with a range of recommendations to lower the barriers and unleash the potential of public procurement for innovation.

Chapter 3, by Ville Valovirta, investigates PPI primarily from an organizational perspective. The author describes how public agencies are faced with the need to develop capabilities in order to manage new organizational processes. The chapter is based on a thorough review of the relevant literature on PPI, studies on public procurement and research on innovation systems. As a result, the chapter identifies a set of management functions required to implement PPI effectively, and derives a list of key capabilities needed for its implementation.

Chapter 4, by Jakob Edler, Max Rolfstam, Lena Tsipouri and Elvira Uyarra, conceptualizes risk and risk management in PPI, and discusses the value of such a conceptualization for PPI practice and policy-making. The chapter first introduces the general concepts of risk, uncertainty and risk management and develops a typology of PPI risks. It also indicates some governance and managerial challenges that the various types of risks pose for PPI. Using the previous typology, the authors illustrate how specific characteristics of the procurement itself, the procurement process and its environment are more likely to be linked to some of the risks identified and not others, and how different situations necessitate different risk management strategies.

Chapter 5, by Hendrik van Meerveld, Joram Nauta and Gaynor Whyles, looks at forward commitment procurement and its effect on perceived risks in PPI projects. Both public sector customers and private sector suppliers perceive risks in procuring innovative solutions. This chapter investigates the effect of applying forward commitment procurement to these risks in order to: (1) assess its capacity for risk management; and (2) support public sector organizations in the improved application of forward commitment procurement. The research uses agency theory and transaction cost economics, explaining the effect of forward commitment procurement on risk in terms of uncertainty and transaction bounded investments. Three case studies are used to show that applying the forward commitment procurement method can indeed contribute to the management of perceived risks.

The chapters included in Part II of the book (Chapters 6–10) reflect upon the lessons from a variety of cases in the United States, China, Greece, the UK and Brazil, respectively. Chapter 6, by Nicholas S. Vonortas, categorizes the main features of PPI in the United States. It details the main stipulations and objectives of the Federal Acquisition Regulation, which is the main set of rules applied by the federal government (and applied, in some slightly adapted form, in many states) in regulating public procurement. As such, this set of rules is instrumental in shaping the incentives and behaviour of public employees regarding procurement. Despite PPI having received much attention in Europe, it is hard to find literature on US PPI practices in cases where public procurement may have promoted innovation outside the national defence and security areas. It is reported that most of the strategic procurement in United States is geared towards achieving social purposes like environmental protection, energy conservation, assisting disadvantaged groups, and so on. However, these are seldom connected to the search of innovative outcomes. The chapter also summarizes several cases of PPI at the federal level.

Chapter 7, by Yanchao Li, Luke Georghiou and John Rigby, examines the application of PPI in the Chinese new energy vehicles programme by using procurement activities carried out by the cities of Jinan and Shenzhen as case studies. It is argued that the programme has stimulated the technological advancement and quantitative growth of new energy vehicles in China despite problems such as regional protectionism and duplicate production. At the city level, motivations for conducting procurement have gone beyond promoting technology diffusion, extending in particular to a desire to stimulate local industrial development. The study concludes that the range of policy instruments employed in this programme (public procurement and subsidies) remains limited. A wider range of favorable framework conditions, such as a fully competitive procurement system, must be in place to achieve the wider objective of nurturing lead markets.

Chapter 8, by Yannis Caloghirou, Aimilia Protogerou and Panagiotis Panaghiotopoulos, explores the application of PPI for e-government services in Greek local authorities. In particular, the study focuses on the Local Government Application Framework pilot project, which was launched by the Central Union of the Greek Municipalities. This project attempts to address the general challenge of providing value-added e-government services, at the same time creating more efficient internal management structures and achieving significant scale economies. In the Greek context, this new procurement approach may be considered as an effort to substitute the usual inefficient and ineffective practices whereby local authorities buy an almost identical information and communication technology package but pay differing licence fees. It is worth noting that these packages are usually turnkey solutions, that is, they are not products typically tailored to the specific requirements of each municipality. In particular, the platform's architectural features support its interconnection and interoperability with other information systems of the public sector in Greece and other EU countries so that cross-sectoral and cross-border services can be provided in the future.

Chapter 9, by Jillian Yeow, Elvira Uyarra and Sally Gee, focuses on the application of PPI to achieve sustainable goals. This chapter uses a single case study to illustrate the procurement of recycled paper by a UK government department. It charts the transformation in procurement from a product to an integrated service, and highlights the procurement of a sustainable innovation to achieve multiple objectives. The study indicates the importance of certain factors for enabling the procurement of a more innovative and sustainable solution. In particular, it illustrates the role of project champions in successful innovative change; a few typical examples of such champions are senior management support, a good working relationship between buyer and supplier, and the creation of a space in which trust and ideas generation can be enabled. Data is drawn from secondary sources, observations and in-depth interviews with public and private stakeholders participating in the process.

Chapter 10, by Cássio Garcia Ribeiro and André Tosi Furtado, reflects upon the application of PPI in an emerging economy. Petrobras, a Brazilian state-owned enterprise, serves as a case study. It is a global leader in the field of deepwater oil production technology and so offers an interesting opportunity to investigate whether public procurement in such a country is used to promote the capability of domestic firms to develop innovations. The authors present the findings of a field survey on P-51, a platform that was ordered by the Brazilian state-owned enterprise and began producing in 2009.

Chapter 11, by Jakob Edler, Charles Edquist, Nicholas S. Vonortas and Jon Mikel Zabala-Iturriagagoitia, concludes the book by summarizing its main lessons and limitations. In this sense, the authors make a clear distinction between PPI as an innovation policy tool, and the practice of PPI to achieve organizational and policy improvements. The chapter closes with a way forward for both academia and policy-makers.

NOTES

1. If a definition of a system of innovation does not include all the determinants of innovation processes, then which of the potential determinants to exclude, and why, have to be justified. Therefore a broad definition seems useful.
2. See, for example, Bergek et al. (2008).
3. This implies that innovation policy also includes public-organization actions that unintentionally affect innovation.
4. In Edquist (2014c), it is also discussed why there is such a large difference between innovation research and innovation policy in this respect.
5. This section is partially based on Edquist and Zabala-Iturriagagoitia (2012, forthcoming) and Edquist (2014a).
6. The public organization may also financially contribute directly to the R&D leading to the development of the product. However, such contributions are not intrinsic parts of the PPI as such. Public R&D funding is a different – complementary – policy instrument in the instrument mix, one which is not in focus here. The purchase of a non-existing product is the central element of PPI. However, the development costs of the new product are, of course, indirectly supported by the procurer by (initially) paying a high price for the product. This is part of the very idea of PPI, but since the procurers' commitment is only to buy a number of units of the product at a certain price, this support of the development cost is brought about through the product price mechanism and cannot be regarded as direct public R&D funding.
7. For a discussion, see Edquist and Zabala-Iturriagagoitia (forthcoming).
8. This is further discussed below, and in Edquist (2014a).
9. See definition of innovation above.
10. For a more thorough analysis of the particularities of pre-commercial procurement, see Edquist and Zabala-Iturriagagoitia (forthcoming).
11. Extensive positive externalities raise the expectation that the innovator

will quickly 'lose' the innovation to competitors (without proper compensation) and thus the ability to benefit privately from it. This expectation, in turn, creates unwillingness to pursue the innovation in the first place, rather to wait for someone else to make the move. Being concerned with social returns, the public sector has exactly the opposite incentives: pursue innovations with high positive externalities.

12. Frank Knight's (1921) differentiation between risk and uncertainty is invoked here. Risk signifies the ability to formulate probability distributions of potential events. This is typically the case with short-term, less radical innovations. Uncertainty signifies the absence of such distributions and is more applicable in the case of longer-term, more radical innovations.
13. There is also a form of procurement called 'forward commitment procurement' (see Chapter 5 in this book). It is actually statements about future procurements. It is not legally binding, but a way to inform potential suppliers about future needs that procuring organizations have.
14. The EU technology platforms or forward commitment procurements may serve as a means of communicating probable future demand.
15. The technology foresight literature has, in fact, produced very valuable justifications and guidance on how to organize such public consultation processes. See, for example, UNIDO (2005).
16. For PPI on the civilian side, it might also be very good experience to learn from the ample experience that has been built up in the procurement of defence equipment. Moreover, there may be retired people with significant experience that public agencies can tap.
17. The acquirer typically does not understand what is possible at a specific point in time. And even if they do understand this to some extent, they do not normally have competence in the field of detailed technical design. The Swedish X2000 high-speed train and the Norwegian public safety network are two examples illustrating how overspecification often leads to the failure of the procurement initiative (Edquist and Zabala-Iturriagagoitia, 2012).
18. The 'industry days' by procuring public agencies in the United States are exactly such an attempt (Edquist and Zabala-Iturriagagoitia, forthcoming).
19. For a review of existing regulations on public procurement in the European context, see the website that the European Commission has created for this purpose, http://ec.europa.eu/internal_market/publicprocurement/index_en.htm (accessed December 2013).
20. Multiple sourcing in the presence of innovative products is the typical approach of industry, for example, in 'private procurement for innovation'. Companies regularly pursue parallel research projects at earlier stages of the pre-innovation process in trying to minimize uncertainties at that relatively inexpensive stage. As new products move into development, prototyping and early production, costs increase exponentially and experimentation tends to be taken out of the system. No technical uncertainties are allowed at the point of scale-up (Fusfeld, 1986).
21. The concept of 'fluid' political needs refers to the short time periods politicians are elected for and the fact that their constituencies evolve. The politicians may therefore try to introduce some of the evolving issues into the project.

22. That innovation policies are, in practice, still dominantly linear and supply-side oriented rather than holistic, systemic and demand-side oriented is shown in Edquist (2014b, 2014c), and as discussed above.

REFERENCES

Aho, E., Cornu, J., Georghiou, L. and Subira, A. (2006). Creating an innovative Europe. Report of the Independent Expert Group on R&D and Innovation appointed following the Hampton Court Summit and chaired by Mr. Esko Aho. European Communities, Brussels, January. EUR 22005.

Bergek, A., Jacobsson, S., Carlsson, B. and Lindmark, S. (2008). Analyzing the functional dynamics of technological innovation systems: a scheme of analysis, *Research Policy*, 37, 407–429.

Borrás, S. and C. Edquist (2013). The choice of innovation policy instruments. *Technological Forecasting and Social Change*, 80(8), 1513–1522.

Dalpé, R. (1994). Effects of government procurement on industrial innovation. *Technology in Society*, 16(1), 65–83.

Edler, J. (2007). Demand based innovation policy. Working Paper 9, Manchester Institute of Innovation Research, Manchester Business School, University of Manchester.

Edler, J. (2013). Review of policy measures to stimulate private demand for innovation: concepts and effects. Manchester Institute of Innovation Research, Manchester Business School, University of Manchester.

Edler, J. and Georghiou, L. (2007). Public procurement and innovation – resurrecting the demand side. *Research Policy*, 36, 949–963.

Edler, J., Georghiou, L., Blond, K. and Uyarra, E. (2012). Evaluating the demand side: new challenges for evaluation. *Research Evaluation*, 21(1), 33–47.

Edler, J., Ruhland, S., Hafner, S., Rigby, J., Georghiou, L., Hommen, L., Rolfstam, M., Edquist, C., Tsipouri, L. and Papadakou, M. (2005). Innovation and public procurement. Review of issues at stake. Study for the European Commission (No ENTR/03/24). Fraunhofer Institute Systems and Innovation Research. December.

Edquist, C. (1997). *Systems of Innovation: Technologies, Institutions and Organizations*. London, Pinter.

Edquist, C. (2005). Systems of innovation: perspectives and challenges, in Fagerberg, J., Mowery, D. and Nelson, R. (eds), *The Oxford Handbook of Innovation*. Oxford: Oxford University Press, pp. 181–208.

Edquist, C. (2009). Public procurement for innovation (PPI) – a pilot study. CIRCLE Paper no. 2009/13, University of Lund.

Edquist, C. (2011). Design of innovation policy through diagnostic

analysis: identification of systemic problems (or failures). *Industrial and Corporate Change*, 20(6), 1725–1756.
Edquist, C. (2014a). Offentlig upphandling och innovation (Public procurement and innovation). A report to Konkurrensverket (Swedish Competition Authority). Uppdragsforskningsrapport 2014:5, May (in Swedish) Available at http://www.kkv.se/upload/Filer/Trycksaker/Rapporter/uppdragsforskning/forsk_rap_2014-5.pdf.
Edquist, C. (2014b). Holistic innovation policy – why, what and how?, Paper presented at the Lundvall Symposium 'Innovation Policy – Can It Make a Difference?', Aalborg, Denmark, 13–14 March.
Edquist, C. (2014c). Efficiency of research and innovation systems for economic growth and employment. Final report from the 2014 ERAC Mutual Learning Seminar on Research and Innovation Policies, European Commission, Brussels, 20 March. (To be published by ERAC, The European Commission, and also at http://charlesedquist.com.) Available at http://www.consilium.europa.eu/media/3210362/erac-mutual-learning-seminarmarch-2014.pdf.
Edquist, C. and Hommen, L. (1999). Systems of innovation: theory and policy from the demand side. *Technology in Society*, 21, 63–79.
Edquist, C., Hommen, L. and Tsipouri, L. (eds) (2000). *Public Technology Procurement and Innovation*. Boston, MA, USA; Dordrecht, Germany; London, UK: Kluwer Academic Publishers.
Edquist, C. and Zabala-Iturriagagoitia, J.M. (2012). Public procurement for innovation as mission-oriented public policy. *Research Policy*, 41(10), 1757–1769.
Edquist, C. and Zabala-Iturriagagoitia, J.M. (forthcoming). Pre-commercial procurement: a demand or supply policy instrument in relation to innovation? *R&D Management*.
European Commission (2007a). A lead market initiative for Europe. COM (2007) 860.
European Commission (2007b). Pre-commercial procurement: driving innovation to ensure sustainable high quality public services in Europe. Brussels, 14.12.2007, COM (2007) 799 final.
European Commission (2011). Proposal for a directive of the European Parliament and of the Council on public procurement. Brussels, 20.12.2011, COM (2011) 896 final.
Expert Group Report (2005). Public procurement for research and innovation. Developing procurement practices favourable to R&D and innovation. DG Research, European Commission. EUR 21793 EN, September.
Flanagan, K., Uyarra, E. and Laranja, M. (2011). Reconceptualising the 'policy mix' for innovation. *Research Policy*, 40(5), 702–713.
French, German, UK Governments (2004). Towards an innovative Europe. A paper by the French, German and UK Governments. 20 February.

Fusfeld, H.I. (1986). *The Technical Enterprise. Present and Future Trends.* Cambridge: Ballinger Press.

Galli, R. and Teubal, M. (1997). Paradigmatic shifts in national innovation systems. In Edquist, C. (ed.), *Systems of Innovation: Growth, Competitiveness and Employment*. London: Pinter, pp. 342–364.

Georghiou, L., Edler, J., Uyarra, E. and Yeow, J. (2013). Policy instruments for public procurement of innovation: choice, design and assessment. *Technological Forecasting and Social Change*, 86, 1–12.

Geroski, P.A. (1990). Innovation, technological opportunity, and market structure. *Oxford Economic Papers*, 42, 586–602.

Godin, B. (2006). The linear model of innovation: the historical construction of an analytical framework. *Science, Technology and Human Values*, 31(6), 639–667.

Guy, K., Boekholt, P., Cunningham, P., Hofer, R., Nauwelaers, C. and Rammer, C. (2009). Designing policy mixes: enhancing innovation system performance and R&D investment levels. The 'Policy Mix' Project. Monitoring and Analysis of Policies and Public Financing Instruments Conducive to Higher Levels of R&D Investments. A study funded by the European Commission-DG Research, March.

Hekkert, M.P., Suurs, R.A.A., Negro, S.O., Kuhlmann, S. and Smits, R.E.H.M. (2007). Functions of innovation systems: a new approach for analyzing technological change. *Technological Forecasting and Social Change*, 74, 413–432.

Hommen, L. and Rolfstam, M. (2009). Public procurement and innovation: towards a taxonomy. *Journal of Public Procurement*, 9(1), 17–56.

IDEA (2013). Bringing research to the market: policy analysis to speed up the market uptake of research results. Draft Interim Report – Background Paper, presented at workshop, DG Enterprise and Industry, Brussels, 17 April.

Izsak, K. and Edler, J. (2011). Trends and challenges in demand-side innovation policies in Europe. Thematic Report 2011 under Specific Contract for the Integration of INNO Policy TrendChart with ERAWATCH (2011–2012). Technopolis, 26 October.

Johnson, A. (2001). Functions in innovation system approaches. Paper for DRUID's Nelso–Winter Conference, Aalborg, Denmark.

Knight, F. (1921). *Risk, Uncertainty and Profit*. Boston, MA: Hart, Schaffner & Marx/Houghton Mifflin Company.

McKelvey, M. (1997). Using evolutionary theory to define systems of innovation. In Edquist, C. (ed.), *Systems of Innovation: Growth, Competitiveness And Employment*. London: Pinter, pp. 200–222.

Mowery, D.C. and Rosenberg, N. (1979). The influence of market demand upon innovation: a critical review of some recent empirical studies. *Research Policy*, 8(2), 102–153.

Nauwelaers, C. (2009). Policy mixes for R&D in Europe. Study Report, European Commission, Directorate-Generale for Research.

Navarro, M. and Magro, E. (2013). Complejidad y coordinación en las estrategias territoriales. Reflexiones desde el caso vasco. *Ekonomiaz*, 83(2), 235–271.
OECD (2011). *Demand-side Innovation Policies*. Paris: OECD Publishing.
Oslo Manual (2005). *Guidelines for Collecting and Interpreting Innovation Data*, 3rd edn. Paris: OECD.
Rothwell, R. and Zegveld, W. (1981). Government regulations and innovation: industrial innovation and public policy. In Rothwell, R. and Zegveld, W. (eds), *Industrial Innovation and Public Policy: Preparing for the 1980s and the 1990s*. London: Pinter Publishers, pp. 116–147.
UNIDO (2005). *Technology Foresight – Manual, Volume 1: Methods and Organization*, Vienna: United Nations Industrial Development Organization.

APPENDIX

Table 1A.1 Key activities in systems of innovation

I. Provision of knowledge inputs to the innovation process
  1. Provision of R&D results, and thus creation of new knowledge, primarily in engineering, medicine and natural sciences.
  2. Competence building, for example, through individual learning (educating and training the labour force for innovation and R&D activities) and organizational learning. This includes formal learning as well as informal learning.

II. Demand-side activities
  3. Formation of new product markets (for example, public procurement for innovation).
  4. Articulation of new product quality requirements emanating from the demand side.

III. Provision of constituents
  5. Creating and changing organizations needed for developing new fields of innovation. Examples include enhancing entrepreneurship to create new firms and intrapreneurship to diversify existing firms, and creating new research organizations, policy organizations, and so on.
  6. Networking through markets and other mechanisms, including interactive learning among different organizations (potentially) involved in the innovation processes. This implies integrating new knowledge elements developed in different spheres of the SI and coming from outside with elements already available in the innovating firms.
  7. Creating and changing institutions – for example, patent laws, tax laws, environment and safety regulations, R&D investment routines, cultural norms, and so on96 – that influence innovating organizations and innovation processes by providing incentives for and removing obstacles to innovation.

IV. Support services for innovating firms
  8. Incubation activities such as providing access to facilities and administrative support for innovating efforts.
  9. Financing of innovation processes and other activities that may facilitate commercialization of knowledge and its adoption.
  10. Provision of consultancy services relevant for innovation processes, for example, technology transfer, commercial information, and legal advice.

Source: Adapted from Edquist (2011).

Table 1A.2 Demand-side policy instruments

Instrument	Method of functioning
1. Public demand: state buys for own use and/or to catalyze private market	
General procurement	State actors consider innovation in general procurement as main criterion (for example, definition of needs, not products, in tenders)
Strategic procurement	State actors specifically demand an already existing innovation in order to accelerate the market introduction and particularly the diffusion
Cooperative and catalytic procurement	State actors deliberately stimulate the development and market introduction of innovations by formulating new, demanding needs (including forward commitment procurement)
	Special form: catalytic procurement: the state does not utilize the innovation itself, but organizes only the private procurement
2. Support for private demand	
Direct support for private demand	
Demand subsidies	The purchase of innovative technologies by consumers or industrial demanders is directly subsidized, lowering the entry cost of an innovation
Tax incentives	Amortization possibilities for certain innovative technologies, in different forms (tax credit, rebate, waiver, and so on)
Indirect support for private demand: information and enabling (soft steering): state mobilizes, informs, connects	
Awareness-building measures	State actors start information campaigns, advertise new solutions, conduct demonstration projects (or support them) and try to create confidence in certain innovations (in the general public, opinion leaders, certain target groups)
Labels or information campaigns	The state supports a coordinated private marketing activity which signals performance and safety features

Table 1A.2 (continued)

Instrument	Method of functioning
Training and further education	Consumers are made aware of innovative possibilities and simultaneously placed in a position to use them
Articulation and foresight	Societal groups and potential consumers are given voice in the market place, signals as to future preferences (and fears) are articulated and signaled to the marketplace. Variations (including constructive technology assessment)
User–producer interaction	State supports firms to include user needs in innovation activity or organizes fora of targeted discourse (innovation platforms, and so on)
Regulation of demand or of the interface demander–producer	
Regulation of product performance and manufacturing	The state sets requirements for the production and introduction of innovations (for example, market approval, recycling requirements). Thus demanders know reliably how certain products perform and how they are manufactured
Regulation of product information	Smart regulation to ensure freedom to choose technologies, but changing the incentive structures for those choices (for example, quota systems)
Process and 'usage' norms	The state creates legal security by setting up clear rules on the use of innovations (for example, electronic signatures)

Support of innovation-friendly private regulation activities	The state stimulates self-regulation (norms, standards) of firms, supports/moderates this process and plays a role as catalyst by using standards
Regulations to create a market	State action creates markets for the consequences of the use of technologies (most strongly through the institutional set up of emission trading), or sets market conditions which intensify the demand for innovations
3. Systemic approaches	
Integrated demand measures	Strategically coordinated measures which combine various demand-side instruments
Integration of demand- and supply-side logic and measures	Combination of supply-side instruments and demand-side impulses for selected technologies or services (including clusters integrating users and supply chains)
	Conditional supporting of user–producer interaction (R&D grants if user involved)
	Pre-commercial procurement

Source: Edler (2013).

PART I

Conceptual framework

2. The meaning and limitations of public procurement for innovation: a supplier's experience

Jakob Edler, Luke Georghiou, Elvira Uyarra and Jillian Yeow

INTRODUCTION[1]

Across the Organisation for Economic Co-operation and Development (OECD) world, public procurement of innovation is becoming an essential element of innovation policy. In turn, this reflects a move towards more demand-based innovation policy to tackle the grand challenges being faced by societies all over the globe (OECD, 2011; Izsak and Edler, 2011). While the economic downturn and the focus on efficiency gains in public expenditure have changed the nature of the debate somewhat, the OECD-wide trend in policy development continues in this direction. It is manifested in specific schemes to push procurement for innovation, and in attempts to improve procurement capabilities and regulations more generally.

However, despite a high level of political intent, there is still a general perception that public procurement has not realized, by far, its potential as an engine for innovation. On the surface, the political climate changed during the economic crisis, and recent years have seen a turn towards more austere public budgets, and with it a shift of the debate on public spending from long-term societal benefit and innovation to reducing costs and securing national supplier benefits.[2] Below that surface there is a range of more basic institutional and procedural reasons for slow progress. Case studies and qualitative analyses cast some light on the range of these reasons. The multilevel nature of public procurement processes in public organizations results in a disconnection between those responsible for concrete procurement and those designing policies mobilizing public procurement for innovation. Further, there are indications that the incentive structures, capabilities and priorities in public organizations are not very conducive to risk-taking, to broad internal and external interaction and to prioritizing long-term delivery plans over short-term budgetary considerations (Rolfstam et al., 2011; Lember et al., 2007; Edler et al., 2005).

To break out of this impasse and to lay the ground for a structural improvement of public sector procurement, we need a better and broader empirical base. More systematic data and generalizable analysis of the meaning of procurement for innovation, and the actual practice of procurement, are still needed. Thus, it is the major aim of this chapter to underpin the current debate on the link between procurement and innovation with empirical findings that complement the valuable insights of the case study work that is, *inter alia*, presented in this book. The analysis presented here is based on a survey of 800 organizations (private firms and third sector) that supply to public sector organizations in the UK at local and national levels across a range of policy areas. This survey, to our knowledge, is the first systematic attempt to analyse the practices and attitudes of suppliers to public bodies with regard to innovation effects of public purchasing.

This chapter first summarizes the main rationales for the link of public procurement with innovation, and turns it into the focus of the policy discourse. It then presents our methodology and sample, and the main analysis which focuses on the innovation activity of the sample. It goes on to analyse the link between innovation and public procurement, the practices used in public procurement, and the perceived barriers. A final section concludes the chapter with some overall discussion and policy recommendations.

PUBLIC PROCUREMENT AND INNOVATION

The connection between public procurement of goods and services and innovation is mainly based on three considerations (Edler and Georghiou, 2007). Firstly, procurement of innovative products and services can have a direct impact on service delivery. It can make public services more effective through improving service delivery or adding new services that add value to citizens; it can also make public services more efficient. Despite the higher search and purchasing costs associated with buying innovations, the adoption and use of an innovative product or service can lead to overall net life-cycle cost savings and thus to overall net benefit over time.

Secondly, public demand for innovation can incentivize industry to invest in innovation, with potentially substantial spillover effects. This can happen in two ways: public procurement can trigger innovations by formulating a new need, and set in motion new innovation cycles; it can also be responsive to novel products and services produced by industry, and thus send a signal to industry that the public sector market is a location in which innovative goods and services can be introduced and diffused. Public demand for innovation can also give suppliers in a given country a leading edge and – depending on the nature of the product or service – potentially initiate further private demand. As public needs are similar in many other countries, innovation procurement can also trigger export opportunities, taking advantage of the UK being a lead market. Furthermore, there is a particular benefit for innovative start-ups. Such firms often struggle to find the first customer to begin their 'reference list'. A public purchase helps to overcome this credibility gap and is worth far more than a grant. On the basis of largely qualitative empirical work, analysts have already characterized procurement policy for many years, as 'a far more efficient instrument to use in stimulating innovation than any of a wide range of frequently used R&D subsidies' (Geroski, 1990, p. 183).

Thirdly, public demand is seen as trigger for products and markets, the diffusion of which helps to support policy goals in other domains, such as energy, health, transport and so on. The car fleets of public organizations or the construction tenders for public buildings can set an example for the private sector, demonstrating feasibility and effectiveness, and thus lowering the entry barriers for innovations in the private sector. This then can multiply the initial public sector purchasing effect beyond its actual spending power. Hence, spillover effects must be taken into consideration when assessing the relative benefit of a public purchase of an innovation.

The effectiveness of procurement in fostering innovation is influenced by the way procurement is undertaken. Public procurement consists of a series of interlinked actions and decisions taken during the procurement life cycle, from the planning stage to the evaluation stage; see Chapter 1 in this volume (OGC, 2004;

Edler et al., 2005). Aspects such as the nature of tender specifications, size of contracts, conditions for participation in tenders, assessment criteria and other elements related to the management of risk and intellectual property rights (IPR) all influence the likelihood of innovation emerging as a result of procurement. For instance, specifications phrased in terms of outcomes or performance are considered more advantageous in allowing industry to propose innovative solutions (Rothwell and Zegveld, 1981; Geroski, 1990). The process of generating and delivering the innovation also depends largely on the ability of buyer and supplier to interact. The Office of Government Commerce (OGC) (2004) suggests the need for early interaction in procurement in order to better 'capture innovation'. If the specifications for a new product or service are done interactively, the likelihood of uptake and use is higher. Further, and more generally, interaction increases the information about costumer needs and preferences (Von Hippel, 1986), but also about supplier abilities and future pipelines.

Besides the potential benefits, much of the debate has centred on potential barriers to innovation in public procurement (Uyarra et al., 2014). Difficulties associated with taking this agenda forward include regulatory complexity, potential conflict between policy objectives, and capacity and resource constraints in contracting authorities (Edler and Uyarra, 2013). Typical barriers and constraints to innovation include lack of interaction between procurers and suppliers, lack of advance communication about potential needs, risk aversion in the granting of contracts, costly and over-bureaucratic tendering procedures, lack of capabilities of procurement procedures, and rigid specifications, among others (OGC, 2004; House of Lords Science and Technology Committee, 2011).

Evidence of the innovation impact of public procurement is rather fragmented and restricted to individual 'success' cases that examine alleged factors contributing to their success, rather than actual impact (Edler et al., 2005; Edquist and Hommen, 2000; Lember et al., 2011). Equally, the empirical evidence on potential barriers is often derived from success cases that show how barriers are overcome in order to make innovation happen (Rolfstam et al., 2011; Lember et al., 2007; Edler et al., 2005). This is because assessing the influence of procurement on innovation is not an easy task. Demand-side policies in general, and public procurement of innovation in particular, present significant challenges for policy evaluation (Edler et al., 2012; OECD, 2011; Uyarra, 2013), which makes 'evidence-based policy making in this area difficult' (OECD, 2011, p. 12).

What actually constitutes the policy can sometimes be difficult to define. As mentioned earlier, procurement cannot be identified with a single instrument but with a series of sub-tasks and mechanisms of a legal, procedural and institutional nature. So recognizing a good procurement system is not a trivial issue. Furthermore, procurement is a multi-objective policy, and it is therefore difficult to disentangle what aspect of procurement has which effect. Finally, when tracing the impact of demand-side measures, it is also difficult to separate their influence on innovation from other policies from the supply-side.

In monitoring terms, difficulties also arise in defining what constitutes innovation in procurement terms, and in relation to measuring the extent and impact of such innovations (Edler and Uyarra, 2013; Uyarra and Flanagan, 2010). Definitional issues arise when trying to define market novelty (according to the Oslo definition of innovation), and when trying to measure the direct and indirect effects of innovation procurement. The frequent lack of reliable administrative procurement data and suitable indicators also poses an important challenge. Key difficulties are associated with adequately categorizing public spending by sector or functional areas, capturing the full extent of procurement activities in a traditionally very fragmented public sector, and identifying the proportion of innovative procurement out of all procurement. Understanding the influence of procurement on innovation in firms is further complicated by the fact that most firms have both government and private sector clients, so it is difficult to separate the effect of public expenditure from overall demand. In the case of organizations such as third sector providers, it is necessary to differentiate the effect of government contracts from other forms of government support.

Quantitative studies assessing the influence of procurement on innovation have typically been of two types: they have either targeted procurement officials, trying to address the extent to which they incorporate innovation consideration in the design of public tenders, and the level of competence and capabilities of the procurement function (CBI/QinetiQ, 2006; Aschhoff and Sofka, 2009; Starzynska and Borowicz, 2012), or directly targeted the suppliers responsible for delivering the contracts. These two types of analysis are often not connected.

Within the latter, studies tend to use innovation surveys to assess the impact of public procurement vis-à-vis (or in combination with) other innovation policy measures on firms' innovation performance (see, for example, Aschhoff and Sofka, 2009; Guerzoni and Raiteri, 2012). While these studies are valuable in demonstrating the effectiveness of procurement in boosting innovation, they are less useful in teasing out the nature of that impact and the precise design of the intervention that is likely to have an impact. In particular, they fall short of unpacking the elements within procurement which are associated with innovation effects, and the main barriers preventing the public sector from realizing this outcome. Innovation surveys also tend to under-survey important areas of public sector service provision (such as social services or health), as well as suppliers such as third sector organizations. On the other hand, dedicated procurement surveys, generally commissioned by the public sector itself (NAO/Audit Commission, 2010) have sought to understand firms' experience of public procurement, but they fail to connect it with improvements in suppliers' competiveness via innovation and improved sales performance. These dedicated surveys are inevitably affected by a selection bias, for they do not represent a random selection of firms, leading to a likely over- or underestimation of the innovation effects of public procurement.

The survey conducted within the UNDERPINN study[3] (Understanding Public Procurement of Innovation) is a first attempt at connecting procurement practices from a supplier perspective with the innovation characteristics and performance of the supplier firms. The characteristics and broad findings of the survey are detailed in the remainder of this chapter, after the UK public procurement context is introduced.

METHODOLOGY: CONTEXT AND DATASET

Public Procurement in the UK[4]

The UK is a good example of a country with high ambition as regards public procurement of innovation. The country has indeed been regarded as a front-runner in pushing the procurement agenda towards innovation (Edler and Uyarra, 2013).

The United Kingdom has a semi-centralized public procurement structure and is governed by the Public Contracts

Regulations 2006 (for England, Wales and Northern Ireland) and the Public Contracts (Scotland) Regulations 2006.[5] The UK public procurement landscape comprises a plethora of organizations performing legislative, audit and improvement roles (for more details see Uyarra et al., 2013). At the national level, the Efficiency and Reform Group (ERG) in the Cabinet Office is tasked with promoting government efficiency and public services reform. The ERG was created in 2010 by putting together expertise from different parts of the Cabinet Office, HM Treasury, Directgov,[6] the Office of Government Commerce (OGC) and Buying Solutions. The National Audit Office (NAO) is in charge of monitoring public spending on behalf of the UK Parliament. The Audit Commission is, in turn, the independent body responsible for ensuring value for money in England in local government. Separate arrangements are in place in Scotland, Wales and Northern Ireland.

Contracting authorities include central government departments and agencies, non-departmental public bodies (NDPBs), devolved administrations, schools, local authorities, housing associations, police forces and NHS trusts. In addition, more than 50 professional buying organizations (PBOs) operate in the UK at the sub-regional, regional and national levels, working along geographical and sectoral lines.

UK public bodies spent about £238 billion in 2010/11 on procurement of goods and services (HM Treasury, 2012). This figure corresponds to 35 per cent of public expenditure on services and about 16 per cent of the UK's gross domestic product (GDP). Local governments are responsible for 33 per cent of the public procurement, while central government, including non-ministerial departments, devolved governments and the National Health Service (NHS) are responsible for the remaining 67 per cent. Within central government, the Department of Health (including the NHS) and the Ministry of Defence account for 58 per cent of central government procurement and 37 per cent of total public procurement (Uyarra et al., 2013). Public procurement of innovation has been high on the political agenda in the UK, particularly since the year 2000, with the launch of a host of initiatives and reports to mobilize the use of procurement to support competitiveness and innovation. The use of public procurement has also been connected to the pursuit of additional policy goals such as sustainability, regeneration and training, although some commentators have noted the danger of excessive fragmentation and potential confusion among procurers (Uyarra and Flanagan, 2010). Specific policy schemes have included the innovation procurement plans (IPPs), the Small Business Research Initiative (SBRI), forward commitment procurement (FCP), a new scheme to link private and public demand (Carbon Compacts), and the Department of Health's Innovative Technology Adoption Programme (iTAPP) (see Uyarra et al., 2014).

The momentum for dedicated innovation procurement policy slowed down in the early 2010s. The Coalition government adopted a different approach towards procurement and innovation, privileging efficiency and fostering initiatives to streamline and centralize public procurement for common goods and services. Thus, the procurement agenda turned towards efficiency and capabilities, with some commentators arguing for a policy that is more focused on UK-based firms as suppliers.[7] A recent report by the House of Lords (House of Lords Science and Technology Committee, 2011) confirmed the importance of using procurement for innovation, but at the same has raised some doubts about the implementation and effects of general innovation procurement practice and dedicated procurement schemes.

The Survey

The UNDERPINN survey seeks to understand suppliers' perception of the effect procurement has on their innovative processes. The survey was conducted using CATI (Computer Aided Telephone Interviewing) during May and July 2011 and addressed general managers or heads of public sector contracts. The questionnaire asked for information on a wide range of issues related to the innovation activities of supplier firms, the types of procurement they are engaged in, as well as general perception of the main practices and competences of procuring organizations, including perceived barriers to innovation.

Under 'public sector' we consider central government departments, local authorities in England and the English National Health Service (NHS). Even if they do not represent the total procurement activity of the UK, their aggregated procurement expenditure nevertheless accounts for over 90 per cent of the total (see Table 2.1). The suppliers to these public sector organizations were identified by matching publicly available

Table 2.1 Breakdown of sample

Type	Categories	Frequency	%
Size of the organization (no. of employees)	Less than 10	82	10.2
	Between 10 and 49	297	37.1
	Between 50 and 250	226	28.2
	More than 250	190	23.8
	Total	795	99.4
Type of organization	Private	649	81.1
	Social enterprise	139	17.4
	Total	788	98.5
How long supplier to the UK public sector	Less than 2 years	4	.5
	Between 2 and 5 years	44	5.5
	More than 5 years	751	93.9
	Total	799	99.9
% of total sales to the UK public sector	Less than 30%	177	22.1
	Between 30% and 60%	222	27.8
	Between 60% and 90%	221	27.6
	More than 90%	165	20.6
	Total	785	98.1
Main category of goods and services supplied	Facilities and management services	91	11.4
	Healthcare equipment, supplies and services	116	14.5
	Office equipment and IT	61	7.6
	Professional services	159	19.9
	Social community care, supplies and services	133	16.6
	Other (incl. utilities, education, transport)	54	6.8
	Construction works	145	18.1
	Total	759	94.9
Main client	NHS	195	24.4
	Local government	423	52.9
	Central government	121	15.1
	Total	739	92.4

information on government contracts[8] for the last available year (2010) with commercial databases (FAME)[9] to obtain the details of 8214 organizations that were suppliers to the public sector.[10] By July 2011, 800 full interviews had been conducted, which represents approximately 10 per cent of the original sample.

Limitations

The survey is not intended to be representative of the overall population of public sector suppliers. Given the way the sampling frame has been constructed, the survey does not reflect procurement activity in the devolved regions of the UK (as it only covers English local government and English NHS) and under-represents certain of the smallest organizations (although micro enterprises are still 10 per cent of the sample). This under-representation of micro-sized firms is a common concern in studies of firm-level innovation processes, leading to a potential overestimation of the levels of innovation, and investment in research and development (R&D) (Freel and Robson, 2004). This, however, should not compromise the usefulness of our findings overall.

BASIC CHARACTERISTICS OF THE SAMPLE

Table 2.1 gives details of the respondents. More than 80 per cent of the respondents are private firms, while 17.4 per cent are social enterprises. Almost half of the sample organizations are small companies (below 50 employees), one-quarter are large (24 per cent of respondents have more than 250 employees), and a slightly larger proportion are medium-sized (between 50 and 249 employees). The organizations span manufacturing activities, but mainly represent service sector activities. They are mainly suppliers of facilities and management services, healthcare equipment, supplies and services, office equipment and information technology (IT), professional services, social community care, supplies and services, construction works and other (including utilities, education, transport). Construction works are represented most (18 per cent), and office equipment and IT least (7.6 per cent). Figure 2.1 gives a further breakdown of firm size by category of goods and services supplied.

At least half of the respondents identified local government as their main public sector clients by volume of contracting; the rest are almost equally distributed between the NHS and central government. There are differences in the firms that supply to different parts of the public sector: suppliers of professional services mainly serve central government, while providers of construction services and social community care mostly supply

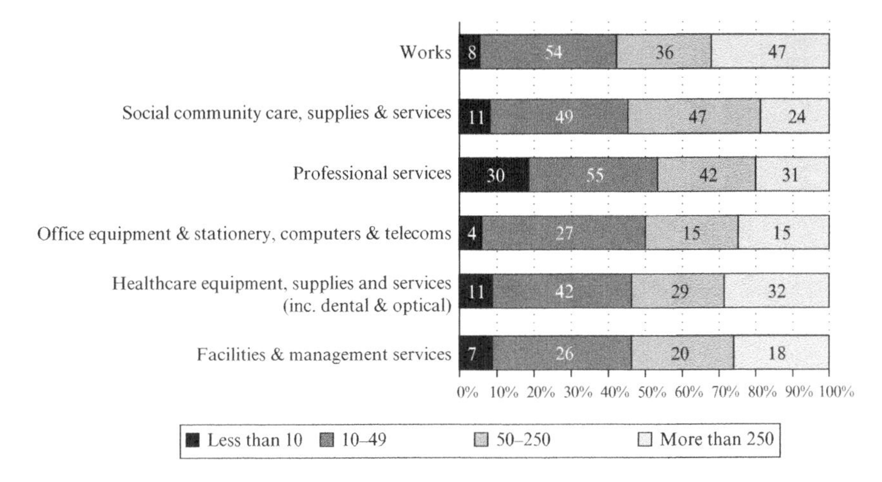

Figure 2.1 Size of firm by procurement category (base: n = 754)

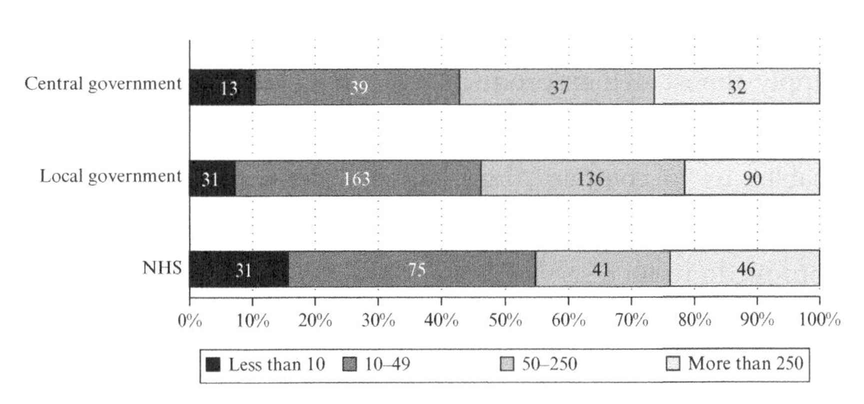

Figure 2.2 Size of firm by area of government (base: n = 735)

to local authorities. Further, suppliers to central government departments are slightly bigger, while there are a higher proportion of smaller suppliers that serve the NHS (Figure 2.2). Finally, there is a stronger presence of third-sector suppliers among local government suppliers compared with central and NHS suppliers.

Respondents tend to be firms that have been suppliers for more than five years (93 per cent), and recognize that they have long-standing relations with the public sector. Importantly for our analysis, there is a spread in terms of their degree of

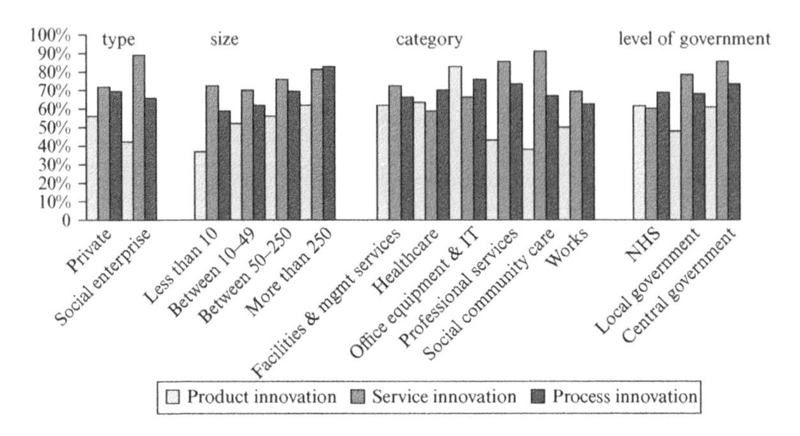

Figure 2.3 Different types of innovation

contracting with the public sector: 22 per cent of the respondents supply mainly to the private sector (that is, the share of public sector costumers is less than 30 per cent), while 28 per cent supply almost all their products and services to the public sector. This means that around 50 per cent of the respondents in the sample serve both public and private sector markets. This also enables us to compare the suppliers' perspective on private and public clients. The vast majority of respondents, around 94 per cent, indicated that they have had an innovation (a new or improved product, service and/or process) within the last three years. More specifically, 54 per cent introduced a product innovation, 75 per cent a service innovation and 67 per cent a process innovation.

Figure 2.3 shows that innovation activities vary among different types of suppliers. Smaller suppliers and suppliers to the NHS are less likely to innovate. Further, most of the innovations of government suppliers are service or process innovations, which is not surprising given the composition of suppliers in the sample in terms of the mix of goods and services and the areas of government they serve. The only exceptions are again suppliers to the NHS, who have a higher share of product innovation; while suppliers for which the dominant market is local tend to be less innovative in terms of products.

Figure 2.4 demonstrates that the public sector plays an important role in the innovation generation process. Two-thirds of the suppliers indicate that the public sector is a very important

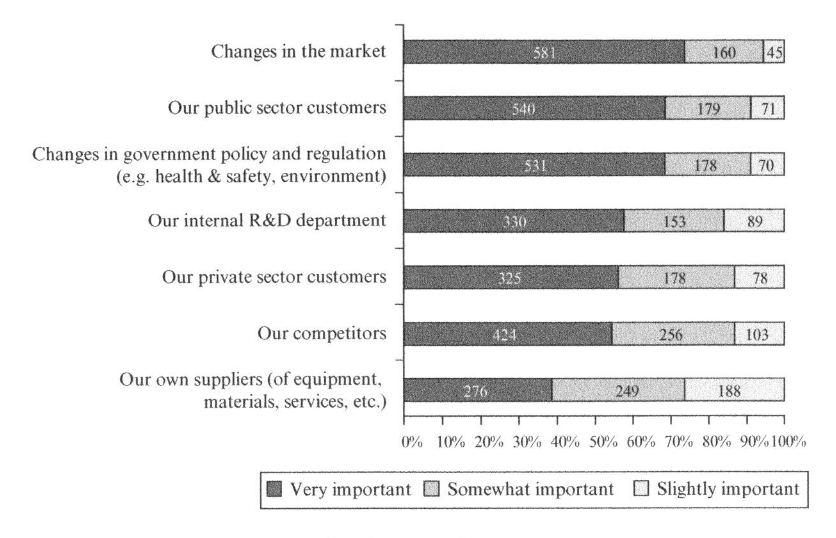

Figure 2.4 Main sources for innovation

source of innovation. This means that, for our sample, the public sector is more important as an innovation source than private costumers, internal R&D or suppliers. This also holds true for the group of suppliers who have roughly equal measures of clients in the private and the public sectors.

Q: How important were the following sources for driving innovation within your organization?

THE EMPIRICAL LINK BETWEEN PUBLIC PROCUREMENT AND INNOVATION

The literature reviewed above suggests a strong link between public procurement and innovation. In this survey we asked suppliers that had introduced a new or improved product or service in the previous three years to confirm whether the innovation had been the result of procurement; namely, if they had innovated in the context (or in anticipation) of bidding for or delivering a contract for the public sector. We also asked them to identify whether they had increased their investment in R&D as a result of involvement in bidding for or delivering a contract. We then tried to identify spillover effects from those innovations initially triggered by public buyers, and asked if they had led

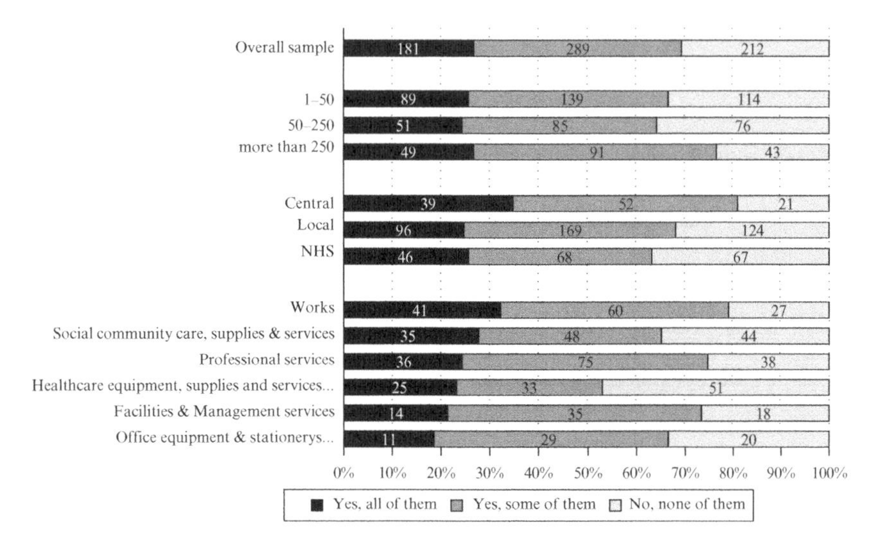

Figure 2.5 Public demand triggering innovation: level of government and size of organization

to additional sales to their public and private sector clients both domestically and internationally (increased exports).

A first major message is that public procurement can indeed be a source of innovation (Figure 2.5). Out of those innovating organizations, 67 per cent indicate that bidding for or delivering contracts to public sector clients has had some impact on their innovation activity; 25 per cent of the innovating organizations claim that all of their innovations have been the result of public procurement. Furthermore, 56 per cent of the sample reports that they won a public sector contract in the last three years because of the innovation-related content in the proposals.

The suppliers that innovate in products are less likely to consider that procurement has had some influence on their innovations compared to those that do not innovate in products. By contrast, suppliers that innovate in services are more likely to consider that procurement has had some influence on their innovations compared to those who do not. Similar responses are provided by suppliers that innovate in processes, although the influence is less clear cut. This suggests that perhaps the public sector is a better place in which to stimulate innovation in services and processes than in products.

In order to analyse the extent to which the public sector demand drives innovation, the suppliers were asked if the innovations suppliers had generated in the last three years were the result of bidding for or delivering to the public sector.

Figure 2.5 shows that, for two-thirds of the sample, government demand has led to an innovation. Suppliers to central government bodies are more likely to report that innovation was a result of procurement when compared with suppliers to the NHS or local government. This may partly be due to the different procurement profile of central government, with a dominance of professional services, but also possibly due to capability constraints affecting local government procurement (see the section on 'Main Barriers to Innovation Procurement') as well as the difficulties associated with procuring innovation in a complex and fragmented NHS (Yeow and Edler, 2012; Phillips et al., 2007; Rosen and Mays, 1998). Equally, large firms are more likely to report the influence of procurement on innovations compared to medium-sized or small firms. Innovations due to bidding or delivering public sector contracts are more frequently reported by suppliers of construction and professional services, and less so by providers of healthcare equipment, supplies and services.

Q: Were innovations in the three years previously the result of bidding for or delivering to the public sector?

While many innovations are not based on R&D, increased R&D nevertheless is one foundation for innovativeness of firms, especially for more radical innovations. To increase R&D is one of the aims of initiatives to spur innovation through public procurement. We can look at those innovations that are R&D-driven in a bit more detail. Of respondents that carried out R&D[11] in the last three years, 51.4 per cent acknowledged an increase in R&D spending because of delivering or bidding for a public sector contract. Again, this share is highest for central government and lowest for local government. Further, out of all R&D-performing organizations, the large ones are more likely than smaller ones to invest more in R&D as a result of procurement. Professional services are the most likely to increase their R&D investments when supplying to or bidding for the public sector.

One major argument for public procurement of innovation is the additional spillover effects in terms of market expansion, including the creation of lead markets (Aschhoff and Sofka, 2009; Georghiou, 2007). More than three-quarters of the suppliers

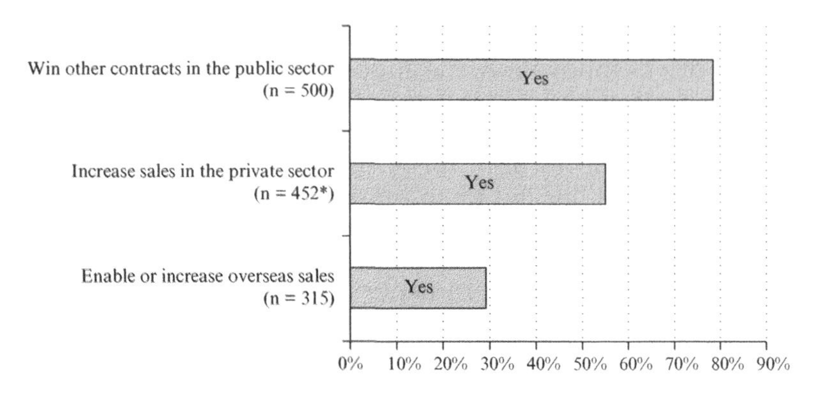

Note: * Excludes those organizations which said that virtually all their sales in the last three years had been to the public sector.

Figure 2.6 The knock-on effects of selling innovation to the public sector

that reported innovations as a result of bidding for or delivering a public sector contract also reported gaining additional public sector contracts as a result of those innovations. More than half experienced increased sales in the private sector, and a smaller share of 29 per cent reported additional overseas sales (Figure 2.6). Our data shows evidence of catalytic effects of public procurement in terms of spillover effects and diffusion in the public sector but also, more importantly, in the private sector at home and abroad, with the potential of generating lead markets (Beise, 2001; Meyer-Krahmer, 2004; Jänicke and Jacob, 2004).

Q: Innovations that resulted from bidding for or delivering public sector contracts have subsequently helped to:

A first, preliminary conclusion out of this first empirical section is that the public sector in the UK is an important source and demander of innovation, which suppliers can benefit in terms of additional sales domestically and abroad. However, the next sections will look in more detail at the practices and modes of procurement enabling this, and at the perceptions of suppliers as to how capable public procurers are at enabling innovation.

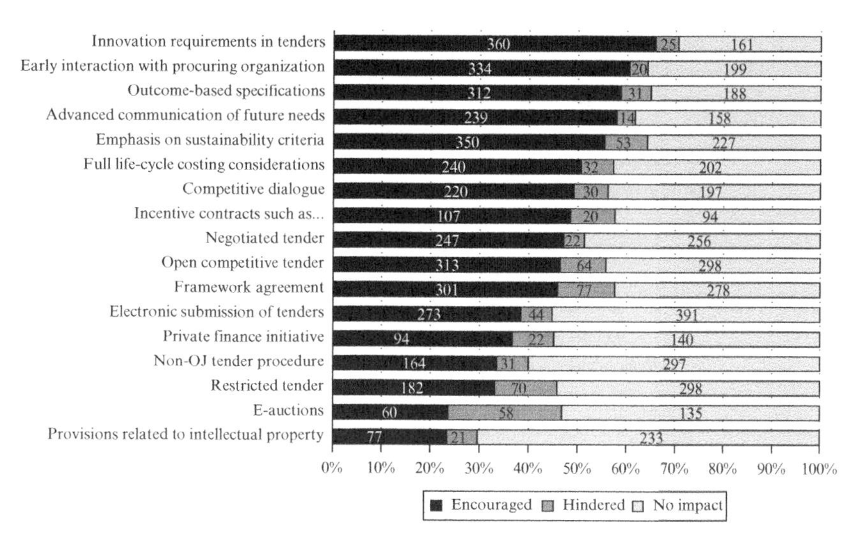

Notes:
Only those respondents who experienced the practice are included.
OJ = *Official Journal* of the European Union.

Figure 2.7 The influence of procurement modes and practices on innovation

MODES AND PRACTICES OF PROCUREMENT FOR INNOVATION

Having established the relative importance of the public sector for innovation, and the basic perception that the public sector could do better, we now turn to the operational level of the procurement process and try to answer the following questions: Which procurement practices and modes are perceived by suppliers as more conducive to innovation? How frequent are these 'innovation-friendly' procurement practices and modes? Underlying these questions is, as mentioned above, the assumption that the different ways in which procurement can be organized influence the likelihood of innovation occurring.

Figure 2.7 shows the influence of procurement modes and practices on innovation. Respondents indicate that certain practices such as innovation requirements in the award criteria of tenders or setting incentives for firms through profit-sharing contracts tend to encourage innovation. Practices also identified

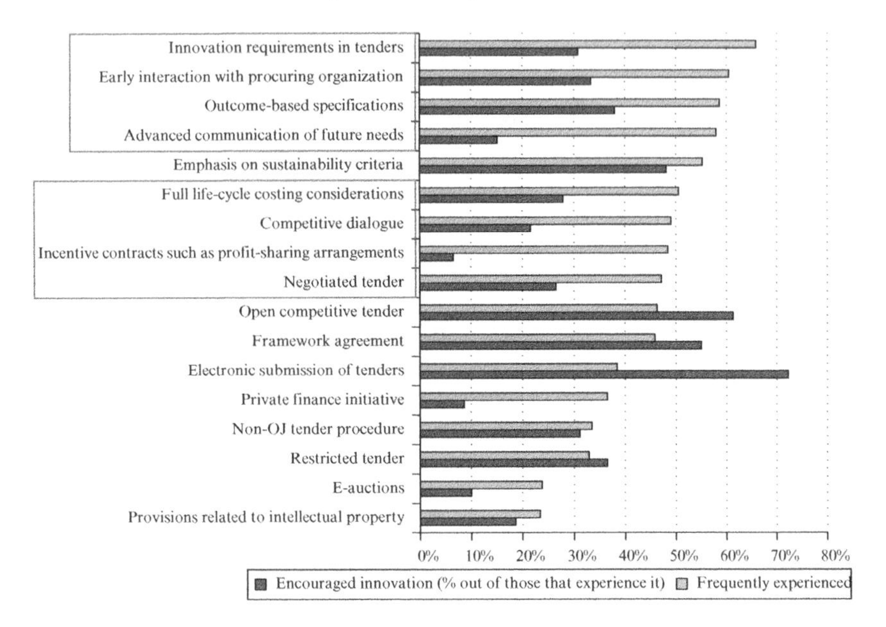

Note: OJ = *Official Journal* of the European Union.

Figure 2.8 The use and meaning of procurement practices

as conducive to innovation include early interaction with suppliers. Pre-procurement communication and early interaction in the process are deemed more important than interactive modes during the process (negotiated tender, competitive dialogue). In general, intelligent or smart procurement practices (outcome specifications, full life-cycle costing, incentive contracts, and so on) are more important for innovation than the choice of modes of procurement (open, restricted). Asking for sustainable products and services appears to be positive for innovation, pointing towards a link between green procurement and the innovation procurement agenda. In contrast, there appears to be a tension between efficiency-oriented processes, such as framework contracts, e-auction or e-tendering, and innovation, as many suppliers see those practices (especially e-auctions) as not fostering or even hindering innovation. Finally, IPR provisions, when set properly, can foster innovation (Figure 2.8), but in general are not seen as used in a particularly innovation-friendly fashion (Figure 2.7).

Q: Has the following encouraged or hindered innovation?

Suppliers were also asked about the procurement practices and procedures they had encountered.

Figure 2.8 turns to this question, and it reveals a central mismatch in the procurement practice. It shows that those procurement practices that foster innovation are not used very often. The gap between the share of suppliers who perceive a practice as encouraging innovation, and the share of suppliers experiencing the practice frequently in Figure 2.8, signals a gap between the innovation-friendliness of practices and the frequency of their use. Innovation requirements and outcome-based specification in tenders as well as early interaction and advanced communication clearly are the most promising processes when it comes to innovation, yet they are not widely used. A further process, likely to lead to innovation but only used very rarely, is the use of incentive contracts such as profit-sharing arrangements. Conversely, practices that are not very innovation-friendly, such as e-tenders, restricted tenders or framework contracts, are used more frequently.

The use of these practices varies across the public sector and across supplier types. Central government suppliers tend to undergo procedures such as competitive dialogue more often than suppliers that serve other parts of the public sector, and also make marginally higher use of framework agreements. Central government is also more likely to include sustainability criteria in its tenders (58 per cent of central government suppliers experience this frequently, vis-à-vis 35.4 per cent of NHS and 49 per cent of local). This is understandable if we consider the type of goods and services procured in different parts of the public sector. Central government is also more likely to include innovation criteria in its tenders (41 per cent of the suppliers admit to experiencing this frequently) and – as indicated above – provisions related to IPR (37.5 per cent).

There are also some differences in the use of practices in relation to different goods and services supplied. For example, competitive dialogue is more frequent in construction and office equipment and relatively rare in social community care. Office equipment and professional services present a higher use of framework agreements. Negotiated tender is particularly common in works and in office equipment. A similar pattern can be identified for the use of restricted tenders, with higher reported use in works, office equipment and professional services. In general, e-auctions are not used much; they are particularly rare in social community care, but more frequent in office supplies and facilities management. Outcome-based specifications are more common in procured services, particularly social services and professional services, where 51 per cent and 42 per cent of suppliers, respectively, have reported experiencing it frequently. E-tendering is widely used across the board, but less so in social community care (17.3 per cent have never used it). The inclusion of sustainability conditions in tenders is something that is experienced frequently in all product types except healthcare, and they are particularly frequent in works where 60 per cent of suppliers have reported experiencing it frequently. Interestingly, there are no major differences in the way suppliers in different product markets have experienced innovation requirements in tenders. IPR is not experienced very frequently (use is particularly low in social care, but also healthcare), and is more commonly used in professional services and office equipment, with 29 per cent and 38 per cent of suppliers, respectively, reporting that they have encountered these practices frequently.

The overall conclusion here is that there is a range of simple practices that are perceived as innovation-friendly by suppliers, but are not rolled out and used broadly. This lack of application of innovation-friendly processes appears to be a major reason for the perceived lack of innovation in public procurement. This leads us to the last step of our enquiry, a deeper analysis of further barriers to innovation in procurement.

MAIN BARRIERS TO INNOVATION PROCUREMENT

As mentioned above, a large number of aspects can prevent suppliers from proposing innovative solutions to tenders, and from increasing their investment in innovation and R&D as a result of procurement. Our survey underpins this debate by asking suppliers about the significance of those barriers to innovation (Figure 2.9) and the competencies of procurers ().

The main barrier is the tendency to give too much weight to price in tenders vis-à-vis quality, as half of the respondents see this as a very significant barrier to innovation and only 15 per cent do not regard it as an obstacle (Figure 2.9). The data shows that the cost issue is much more important as a barrier than a general lack of demand for innovation. The austerity budget and the turn towards efficiency savings continue to work in this direction.

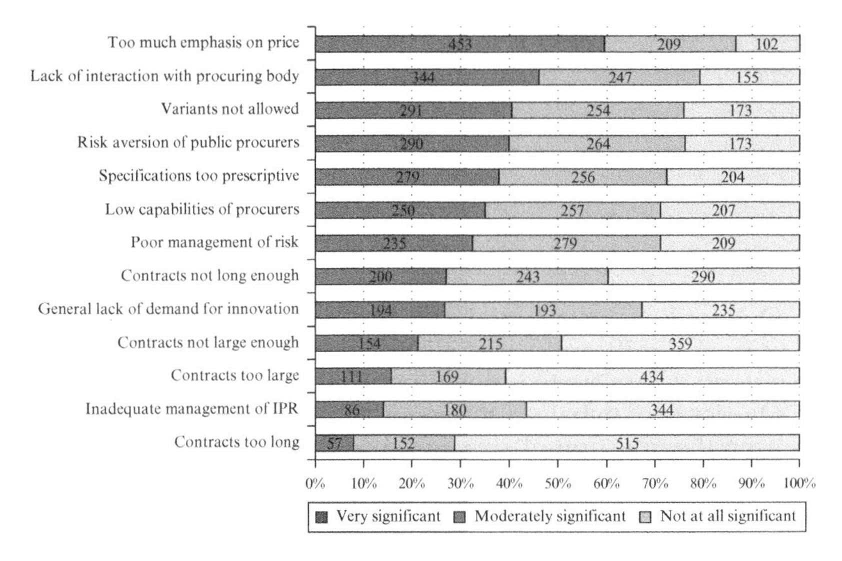

Figure 2.9 Significance of barriers

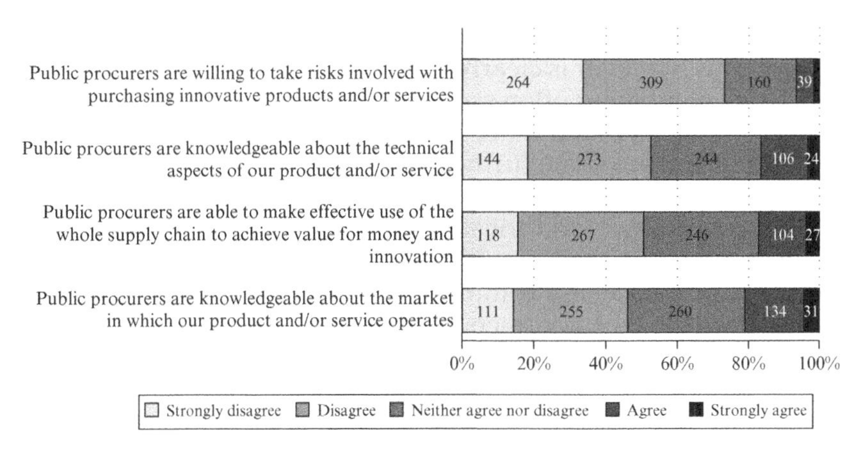

Figure 2.10 Assessment of procurer characteristics

The second most important bundle of barriers has to do with restraining variety: innovation is about variety, about trial and error, and choice. The disallowance of variants, specifications that are very prescriptive and the lack of openness to unsolicited ideas are all practices that close down on variety at a very early stage, and thus are hindrances to innovation. This limitation is linked to perceived risk aversion, poor risk management and a lack of procurer's capabilities as regards market and technological knowledge tapping into supply chains.

Supporting the findings on interaction above, suppliers see a lack of interaction with the procuring body as a key obstacle to innovation. This is true despite the fact that our sample is often characterized by very long-lasting relationships. This indicates that long-lasting buyer–supplier relationships in themselves do not support innovation activity; what is important is interaction to communicate needs.

Q: Based on your experience of public procurement, please tell us how significantly, if at all, you perceive any of the following potential barriers to innovation to be.

One issue of debate in the innovation procurement literature has been the question of size and duration of contract, whereby larger and longer contracts could work both for innovation (incentive to invest) and against innovation (higher risks involved, capacity issues for small, innovative companies) (Cabral et al., 2006; GHK, 2010). Indeed there is a split in the sample in relation to these questions, with a slight majority of suppliers indicating that longer and bigger contracts are more likely to lead to innovation. This however is clearly related to size; the analysis shows that small companies often see contracts as being too large, while large companies much more often think contracts are not large enough. There are some further noteworthy differences between different kinds of companies. Small companies and suppliers of healthcare products and construction works see the emphasis on price and the prescription of tender specification as particularly problematic. Social and community care organizations complain about the fact that contracts are not long enough. Moreover, large companies emphasize the lack of risk management in the public sector as a specific hindrance depicts suppliers' assessment of the attitude and capabilities of public procurers. It shows that more than half of the sample regards the public sector as being risk-averse, showing a lack of knowledge about markets and technological aspects of their products, and being unable to make proper use of the supply chain.

These assessments of the public sector are further pronounced when comparing the perceptions of suppliers about private and public customers. Figure 2.11 shows this comparison, whereby only those suppliers who supply to both private and public customers were included. Those firms consider public buyers to be less innovation-friendly than private customers,

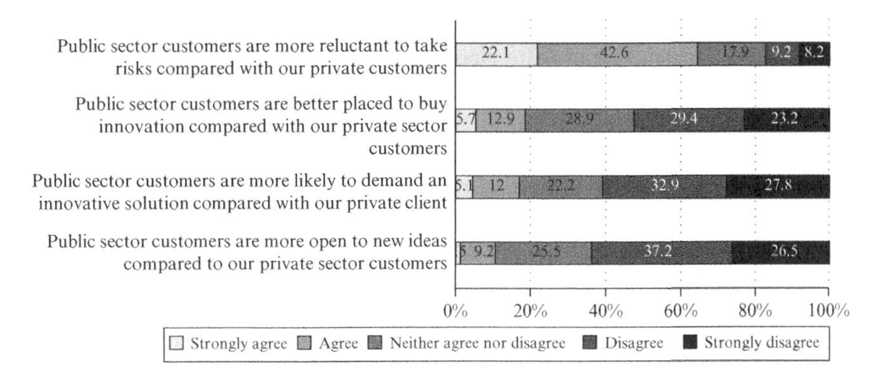

Note: The figure shows answers for those firms and organizations that supply 30–60 per cent of their products and services to public firms, as they are best placed to compare (N varies between 194 and 200).

Figure 2.11 Assessment of public versus private customers

that is, less open to new ideas, less well placed to buy an innovation, less willing to take risks (see Chapter 4 in this book) and less likely to demand innovation in the first place. This is interesting if we consider that the public sector is an important source for innovation for at least two-thirds of respondents. This suggests that even though public sector organizations are key sources for driving innovation within the supplying firms and third-sector organizations, this potential is not fully exploited.

In sum, our survey indicates a broad range of barriers that hinder innovation in public procurement. Those barriers have to do with the lack of risk-taking and the lack of knowledge of and openness to innovation by those actors in public organizations who procure. These findings for the first time confirm the anecdotal evidence in case studies (Rolfstam et al., 2011; Lember et al., 2007; Edler et al., 2005) on the basis of a broad quantitative survey.

MAIN CONCLUSIONS AND POLICY ISSUES FOR DEBATE

This chapter presents the results of the first survey of suppliers to the public sector that is entirely focused on the link of public procurement practices to innovation. This research builds on and complements the existing largely qualitative analysis of innovation in the context of public procurement. While the broad canon of existing case-study work allows us to understand the importance of institutional contexts, structures and practices in forensic detail, it is always limited to specific situations, to selected public organizations and to selected markets. This survey enables us to assess the bigger picture, to understand the relative importance of public procurement for innovation and, more importantly, the enablers of and barriers to innovation as perceived by organizations supplying the public sector. The supplier perspective in the survey is crucial, as it is suppliers who react to signals, who base their decisions to innovate on their interpretations of practices and structures in public procurement. While the survey has its limitations, in that it is focused on one national context, and that the underlying contract data which has helped us to construct the population and sample is not as comprehensive as one would like, it nevertheless allows us to draw some bold and general conclusions on the link between public procurement and innovation.

Our analysis gives clear indications as to the potential of public procurement to drive innovation in the UK, the scope and nature of this innovation, and the barriers inhibiting this potential to be realized to a greater extent. More than half of the sample claim that the fact that they could offer an innovative product or service has won them a public contract in the last three years; public buyers respond to innovation and do trigger innovation, and they are an important source of innovation – more important than private customers or suppliers. Much of this innovation is hidden, in service provision or in processes that might not be apparent to the user or citizen. Further, to supply innovation to public bodies has a catalytic effect: it triggers further innovation and economic effects in the private and the public sector. Public bodies are important first customers and in many cases (one-third of exporting companies) help to increase exports.

Overall, large firms tend to be able to generate and sell innovation for and to public clients slightly more easily than medium-sized and smaller companies do, but the disadvantage of smallness is limited. In general, the central-level procurement tends to be slightly more prone to buy in innovation. This tendency, however, is linked to the nature of the services and goods demanded (professional services largely).

There are many preconditions required to make innovation

a feature of public procurement. The most important one is simple: there must be a commitment to innovation. Innovation must be asked for or triggered explicitly, by allowing variety, by assessing tenders not solely on price and by accepting risk. Innovation needs variety; it needs a signal that variety is allowed and can be rewarded. At the same time, suppliers need some orientation that directs that variety towards future needs. Early signals and pre-procurement interaction appear to be of paramount importance for suppliers and the confidence they have to take the risk of innovating when supplying to the public sector. The timing is important here, since the late processes that spur interaction during the actual tender or negotiation procedures are less important for innovation than processes during the early, more informal stages.

Thus, there is a clear need for public procurement processes to be understood more broadly, and to be extended much more systematically towards pre-procurement activities. This requires public bodies to communicate and signal future innovation needs as early as possible, and to be more explicit about innovation needs. Further, procurement should implement much more broadly what innovation procurement handbooks have been suggesting for years, that is, enabling variety, focusing on outcome and on whole life-cycle costing. The trend towards more standardization, framework contracts and efficiency-oriented tender processes thus represents a clear challenge for the innovation agenda.

Despite the level of innovation that suppliers realize when supplying the public sector, more innovation appears to be possible. There are manifold barriers in the system that hinder a broader demand for, and roll-out of, innovation in public procurement. The assessment made by suppliers of the procurer's capabilities and practices is generally poor, and certainly poorer than that for the private sector. While the risk aversion of public procurers indicated in the survey confirms case study evidence, the lack of market and technological knowledge and the consequences for the lack of innovation are stronger than expected. Importantly, there is no one lever across all company sizes or product categories. Our data shows some influences and barriers that are common across size, sector and type of organization, and others that are specific to these dimensions.

The results of the survey should lead to a more informed discussion on the design and roll-out of public policy measures to improve the capabilities and willingness of the public sector to ask for and adopt innovation. We know from analyses of public policy (Georghiou et al., 2013) that there is a range of policy measures already developed that suffer from a sustained roll-out. The analysis presented in this chapter highlights the areas on which the design and roll-out of support should focus. Public policy schemes should focus on encouraging and enabling procurers and decision-makers to use those modes and practices that are more conducive to innovation. They should put more emphasis on making those responsible for defining needs and organizing the procurement process in public organizations aware of the long-term benefit of innovation, and of the importance of signalling their readiness to absorb innovation. Public organizations need support to establish a culture which sends out those signals both in informal interactions independent of concrete procurement processes, and in the concrete practices of procurement. This also involves making provisions within the public sector to reduce risk aversion and improve risk management. Finally, the perceived lack of technical and market intelligence points towards the need for policy to improve capabilities within the procurement process more generally. Especially in cases in which the public body seeks to trigger a specific innovation to meet its specific needs, this could be supported by a twofold strategy: first, those defining the need within public bodies should drive the procurement process, as they hold the critical market and technology knowledge within organizations (the users, the business case holders); second, the buying organizations could be helped to understand the market and clearly define the need. This can be done via two-step procurement processes (pre-commercial or design contests). These processes could be supported by specialized public organizations but owned by the buying organization, with the aim of building up procurement expertise across the public sector more broadly.

ACKNOWLEDGEMENTS

This work is based on the UNDERPINN study that was funded through the UK Economic and Social Research Council Grant [RES-598–25–0037] with contributed support from the Department of Business Innovation and Skills, NESTA (National Endowment for Science, Technology and the Arts) and the

Technology Strategy Board. The authors would like to acknowledge their generous support and their overall comments on our work. However, no representative of the funding bodies had any influence on the content of this chapter or the decision to publish. The authors would also like to thank Colin Cram, Kieron Flanagan and Sally Gee for their feedback on the survey, and the firms that participated in the survey.

NOTES

1. This chapter is based on a background paper produced in the context of the ESRC funded project UNDERPINN by Edler et al. (2011).
2. See debate in *Daily Telegraph*, 11 September 2012; http://www.telegraph.co.uk/finance/financialcrisis/9534993/Buy-British-and-don't-worry-about-Europe-says-Vince-Cable.html.
3. For further information see https://underpinn.portals.mbs.ac.uk/.
4. This section is partly based on Uyarra et al. (2013).
5. These Regulations implement into UK law the European Commission's Directive on public procurement (2004/18/EC), adopted in March 2004.
6. Directgov, later replaced by Gov.UK, is the main website of all UK government services and departments.
7. See note 2.
8. Overall, we gathered information on public sector contracts for 93 local authorities in England. This is around a third of local authorities in England (not biased towards any one type of authority), the totality of central government departments, and spending for five (South East, East of England, East Midlands, North West, Yorkshire and Humber) of the nine English NHS regions.
9. See http://www.bvdinfo.com/en-gb/products/company-information/national/fame.
10. Only 'core suppliers' to the public sector were considered, namely whose aggregated annual contracting with the public sector was above a set threshold of £50 000 in the 2010 financial year.
11. In our survey, R&D refers to any activities undertaken to increase knowledge for innovation. Examples include making prototypes, testing of a new design, developing new software or IT tools, and conducting market research.

REFERENCES

Aschhoff, B. and Sofka, W. (2009). Innovation on demand – can public procurement drive market success of innovations? *Research Policy*, 38, 1235–1247.

Beise, M. (2001). *Lead Markets: Country Specific Success Factors of the Global Diffusion of Innovations*. Physica Verlag: Heidelberg.

Cabral, L., Cozzi, G., Denicolo, V., Spagnolo, G. and Zanza, M. (2006). Procuring innovations. In Dimitri, N., Piga, G. and Spagnolo, G. (eds), *Handbook Of Procurement*. Cambridge: Cambridge University Press.

CBI/Qinetiq (2006). *Innovation And Public Procurement – A New Approach To Stimulating Innovation*. London: CBI.

Edler, J. and Georghiou, L. (2007). Public procurement and innovation: resurrecting the demand side. *Research Policy*, 36, 949–963.

Edler, J., Georghiou, L., Blind, K. and Uyarra, E. (2012). Evaluating the demand side: new challenges for evaluation. *Research Evaluation*, 21, 33–47.

Edler, J., Georghiou, L., Uyarra, E. and Yeow, J. (2011). Procurement and innovation: underpinning the debate – background paper. Manchester: Manchester Business School.

Edler, J., Ruhland, S., Hafner, S., Rigby, J., Georghiou, L., Hommen, L., Rolfstam, M., Edquist, C., Tsipouri, L. and Papadokou, M. (2005). Innovation and public procurement. Review of issues at stake, study for the European Commission, Final Report. Brussels: European Commission.

Edler, J. and Uyarra, E. (2013). Public procurement of innovation. In Brown, S.O.L. (ed.), *Handbook of Innovation in Public Services*. Cheltenham, UK and Northampton, MA, USA: Edward Elgar.

Edquist, C. and Hommen, L. (2000). Public technology procurement and innovation theory. In Edquist, C., Hommen, L. and Tsipouri, L. (eds), *Public Technology Procurement And Innovation*. Boston, MA: Kluwer Academic Publishers.

Freel, M.S. and Robson, P.J.A. (2004). Small firm innovation, growth and performance: evidence from Scotland and Northern England. *International Small Business Journal*, 22, 561–575.

Georghiou, L. (2007). Demanding innovation lead markets, public procurement and innovation. Nesta Provocation 02. London. Available at http://www.nesta.org.uk/sites/default/files/demanding_innovation.pdf.

Georghiou, L., Edler, J., Uyarra, E. and Yeow, J. (2013). Policy Instruments for public procurement of innovation: choice, design and assessment. *Technological Forecasting and Social Change*, 86, 1–12.

Geroski, P.A. 1990. Procurement policy as a tool of industrial policy. *International Review of Applied Economics*, 4, 182–198.

GHK (2010). Evaluation of SMEs' access to public procurement markets in the EU. Brussels: DG Enterprise and Industry.

Guerzoni, M. and Raiteri, E. (2012). Innovative procurement and R&D subsidies: confounding effect and new empirical evidence on

technological policies in a quasi-experimental setting. Universita Di Torino, Department Of Economics Working Paper Series.
HM Treasury (2012). *Public Expenditure Statistical Analyses 2012*. London: HM Treasury. http://www.hm-treasury.gov.uk/d/pesa_complete_2012.pdf.
House of Lords Science and Technology Committee (2011). *Public Procurement As A Tool To Stimulate Innovation*. London: Stationery Office.
Izsak, K. and Edler, J. (2011). Trends and challenges in demand-side innovation policies in Europe. Thematic Report 2011 under specific contract for the integration of Inno Policy TrendChart with Erawatch (2011–2012). Brussels.
Jänicke, M. and Jacob, K. (2004). Lead markets for environmental innovations: a new role for the nation state. *Global Environmental Politics*, 4, 29–46.
Lember, V., Kalvet, T. and Kattel, R. (2011). Urban competitiveness and public procurement for innovation. *Urban Studies*, 48, 1373–1395.
Lember, V., Kalvet, T., Kattel, R., Penna, C. and Suurna, M. (2007). Public procurement for innovation in Baltic metropolises – case studies. Tallinn University of Technology, Tallinn. Available at http://baltmet-inno.latreg.lv/uploads/filedir/File/BM%20Inno%20Procurement%20for%20Innovation.pdf.
Meyer-Krahmer, F. (2004). Vorreiter-Märkte Und Innovation. Ein Neuer Ansatz Der Technologie Und Innovationspolitik. Hamburg.
NAO/Audit Commission (2010). *A Review of Collaborative Procurement across the Public Sector*. London: National Audit Office.
OECD (2011). *Demand Side Innovation Policy*. Paris: OECD.
OGC (2004). Capturing innovation: nurturing suppliers' ideas in the public sector. London: UK Office of Government Commerce.
Phillips, W., Knight, L., Caldwell, N. and Warrington, J. (2007). Policy through procurement – the introduction of digital signal process (DSP) hearing aids into the English NHS. *Health Policy*, 80, 77–85.
Rolfstam, M., Phillips, W. and Bakker, E. (2011). Public procurement of innovations, diffusion and endogenous institutions. *International Journal of Public Sector Management*, 24, 452–468.
Rosen, R. and Mays, N. (1998). The impact of the UK NHS purchaser–provider split on the 'rational' introduction of new medical technologies. *Health Policy*, 43, 103–123.
Rothwell, R. and Zegveld, W. (1981). *Industrial Innovation and Public Policy: Preparing for the 1980s and the 1990s*, Westport, CT: Greenwood Press.
Starzynska, W. and Borowicz, A. (2012). The pro-innovative effect of Poland's public procurement system on the country's economy. Paper presented at the UNDERPINN Conference on 'Demand, Innovation and Policy. Underpinning Policy Trends with Academic

Analysis', Manchester Institute of Innovation Research, Manchester Business School, University of Manchester, 22–23rd March.

Uyarra, E. (2013). Review of measures in support of public procurement of innovation. Report within the Mioir-Nesta Compendium of Evidence on Innovation Policy. London and Manchester.

Uyarra, E., Edler, J., Garcia-Estevez, J., Yeow, J. and Georghiou, L. (2014). Barriers to innovation through public procurement: a supplier perspective. *Technovations*, 34(10), 631–645.

Uyarra, E., Edler, J., Gee, S., Georghiou, L. and Yeow, J. (2013). UK public procurement of innovation: the UK Case. In Lember, V., Kattel, R. and Kalvet, T. (eds), *Public Procurement, Innovation and Policy: International Perspectives*. London: Springer-Verlag.

Uyarra, E. and Flanagan, K. (2010). Understanding the innovation impacts of public procurement. *European Planning Studies*, 18, 123–143.

Von Hippel, E. (1986). Lead users: a source of novel product concepts. *Management Science*, 32, 791–805.

Yeow, J. and Edler, J. (2012). Innovation procurement as projects. *Journal of Public Procurement*, 12, 488–520.

3. Building capability for public procurement of innovation

Ville Valovirta

INTRODUCTION

As governments are adopting public procurement for innovation as a focal instrument in their toolboxes for demand-driven innovation policies, public sector organizations are faced with a need to develop capability to manage new organizational processes. Public procurement for innovation (PPI) is a policy instrument which requires adopting novel skills and management practices. Innovation policy-related efforts to use public procurement to promote supplier innovation will encounter intersecting goals at the level of public agencies, and are not likely to be translated into effective implementation without building required capabilities. Indeed, recent research has indicated that insufficient management skills have accounted for failures in PPI projects (for example, Rolfstam, 2007). Many of these capabilities are not part of regular procurement competencies. General management also needs to adjust practices to reap the full benefits of the novel approach, and new organizational practices extending beyond the procurement function may need to be established.

Market-based governance models and increased use of outsourcing and contracting out service production have transformed many public agencies from service production organizations to procurement organizations. Purchasing and contracting have become core functions in many public agencies. This development, together with the increasing complexity of various technical systems, has led public agencies to purchase larger and more complex objects. The emphasis has shifted from procurement of goods to service contracting and purchase of complex service–product combinations (Schapper et al., 2006; Caldwell and Howard, 2011). As digitalization of government activities advances, various types of integration issues also emerge. The shift towards more complex supply chains requires understanding of value chain structures and industry capabilities (Schapper et al., 2006). Advanced management capability is needed on the buyer's side to understand suppliers' potential to respond to demand signals for new innovations along the value chain.

Two kinds of impacts are expected from PPI. On the one hand, demand for new or improved goods and services is expected to stimulate innovation among firms supplying to public organizations. Impacts on firm innovation are pursued by innovation policy-makers aiming to exploit the demand power of government purchasing and to complement supply-side support measures with demand-side policy instruments. On the other hand, innovative goods and services, once delivered to the purchasing organization, are expected to improve the quality, productivity and effectiveness of public services, or respond to the anticipated future needs of the society. Public service improvement is sought after by inducing organizations to develop their service delivery and attain policy domain goals. The need for these improvements is driven by multiple factors, such as global socio-economic challenges, austerity of public finances and increased citizen expectations of service quality, exerting pressure on governments to improve service delivery.

The majority of scholarly writing has investigated PPI from a policy perspective to measure its economic impacts (for example, Dalpe et al., 1992; Dalpe, 1994; Aschhoff and Sofka, 2009), and to examine its rationale as an instrument of demand-side innovation policy (Edquist and Hommen, 1999; Edler and Georghiou, 2007; Edquist, 2011; Edquist and Zabala-Iturriagagoitia, 2012). Less attention has been given to managerial issues that need to be addressed when a public organization aims to implement PPI and accommodate it as a regular organizational practice.

Managerial issues have been most notably discussed in recent studies focusing on risk management in innovation procurement (Kalvet and Lember, 2010; Tsipouri et al., 2010; Edler et al., Chapter 4 in this volume). The management perspective, together with a focus on innovation, is also debated in recent work on procurement of complex performance (Caldwell and Howard, 2011). While these studies on risk management and complex performance are highly relevant for the topic at hand, they touch only upon specific aspects (risks, complexity) of managing PPI. This chapter seeks to address the gap in the literature covering broader perspectives related to managing PPI.

As the scholarly discussion has primarily focused on PPI as an innovation policy tool, the service improvement goals of procuring organizations have been given less attention. I argue that, in order to manage PPI effectively, organizations need to embed required capabilities more firmly within an organizational context. Goals related to innovation policy will essentially play a secondary role, subordinate to service improvement goals. We also claim that these two concurrent aspects, while necessary for PPI to take place, make it a domain of multiple and even diverging expectations, which complicate managing the process in a linear manner.

This chapter investigates PPI primarily from an organizational perspective. We review the extant literature and derive a list of key capabilities needed for implementation. Due to scant attention to managerial issues, it is necessary to review the literature on public procurement outside the scope of innovation research literature as well. This provides useful insights into various relevant aspects such as goals set for PPI, market interaction, specification of procurement needs, risk management and coordination issues. The focus is on management aspects specific to PPI as contrasted with regular public procurement, and, to some extent, with other innovation policy instruments.

CAPABILITIES FOR PUBLIC PROCUREMENT OF INNOVATION

Managing the attainment of organizational goals requires building organizational capability and establishing appropriate routines. Winter (2000) defines organizational capability as 'a high-level routine (or collection of routines) that, together with its implementing input flows, confers upon an organization's management a set of decision options for producing significant outputs of a particular type' (p. 991). This definition highlights behaviour that is the result of learning, creates patterns through repetition, and is founded partly in tacit knowledge. Organizational improvisation is thus not a routine. Organizational routines are repetitive, recognizable patterns of interdependent actions, carried out by multiple actors (Feldman and Pentland, 2003) that constitute organizational capabilities (Nelson and Winter, 1982; Becker et al., 2005).

Managing PPI, like any organizational function, requires setting in place necessary routines and practices which in aggregate constitute organizational capability. While actual routines in specific organizations will take a variety of forms, a collection of general management functions can be derived from the research literature. The first step to take is to identify management functions associated with implementing PPI. These functions partly pertain to the functions of innovation systems as specified by Hekkert et al. (2007), partly to the functions that serve principal service delivery missions (for example, public health, safety, or public transportation) and that public procurer organizations pursue, and partly to functions assigned to public procurement in general (Schapper et al., 2006). This multifold target setting is a fundamental feature impacting upon management of PPI by generating a context with a certain degree of complexity.

In the next sections I identify the principal functions of organizational capability to effectively implement PPI. The functions have been identified through a review of the relevant literature on PPI, studies on public procurement, and research on innovation systems.

Setting Goals for Public Procurement for Innovation

Drawing on Winter (2000), I conceptualize organizational capability as a collection of routines to produce significant outputs of a certain type. The outputs that PPI is expected to produce do not yet exist and need to be developed by the supplier; that is, innovation. It is argued that two coinciding benefits are delivered. First, PPI is expected to contribute to solving societal challenges and improving the delivery of public services by sourcing better products and services from suppliers. Second, government demand for new solutions provides opportunities for supplier firms to develop new products to meet these needs. Providing the initial demand and first references helps companies to commercialize their products to other clients, thus generating economic benefits through a diffusion of innovation. Using public demand to spur innovation holds thus a twofold promise to improve public services, while at the same time making a contribution to innovation dynamics in the economy (Edler et al., 2005; Edquist and Zabala-Iturriagagoitia, 2012).

When innovation generation (Edler and Georghiou, 2007) is set as an expressed goal for public procurement, it adds to the list of other objectives set for public procurement. According to Schapper et al. (2006), public procurement aims to achieve three principal objectives: first, public confidence in the fairness of the procurement process; second, efficiency and effectiveness of the money being spent; and third, consistency with various other policies. These goals are not always perfectly aligned in procurement practice, and various stakeholders tend to value them differently.

Introducing promotion of innovation as an expressed goal for public procurement shifts this balance in ways which have also managerial implications. Most obviously, PPI has the potential to contribute to improved long-term efficiency and effectiveness of public services by encouraging firms to develop better solutions to government needs. PPI has the potential to contribute both to incremental improvements and to the radical increase of value for money in public spending.

Moreover, public procurement is already subjected to other socio-economic expectations parallel to innovation targets. These 'horizontal' policy goals include environmental objectives, economic development of disadvantaged groups, employment and promotion of small businesses (Piga and Thai, 2007; Arrowsmith, 2010). Public procurement is thus widely used as a policy tool to advance various socio-economic goals. Public procurement accommodates multiple stakeholder objectives, which can be in contradiction with each other (Caldwell et al., 2005; Arrowsmith, 2010). Although other socio-economic goals are partly overlapping with objectives of PPI, the existence of multiple parallel goals of public procurement adds an element of complexity to managing PPI. PPI can hardly occupy a preferential position with regard to other legitimate socio-economic goals.

It has also been argued that, while setting social and economic goals to procurement is legitimate, these goals should primarily be pursued with policy instruments other than procurement (Eliasson, 2010; Uyarra and Flanagan, 2010). This implies that the generation of innovation, in effect, needs to remain a secondary objective compared to mainstream goals of public procurement; they should be indirect interventions focusing first on the quality of government services, which in turn will indirectly promote innovation in the industry (Dalpe, 1994).

As compared to supply-side support measures, such as

research and development (R&D) grants, which are measures specifically targeted to support innovation, PPI needs to simultaneously serve multiple goals and meet a variety of expectations. The ability to balance innovation goals with other procurement objectives thus appears as an important dimension in the management of PPI. Government agencies also have a variety of missions: an agency providing public services attends primarily to the quality and productivity of their services, whereas innovation policy agencies have a mission to support economic growth through industrial innovation and diffusion. Government is not a monolithic corporation but consists of a variety of sectors and administrative levels, each with specific missions. A capability is required to integrate the innovation generation rationale within sectoral policy rationales, and coordinate efforts to create interorganizational win–win situations (Edler and Georghiou, 2007).

Capability 1: To integrate and balance innovation goals with other procurement objectives within the government and to create interorganizational win–win situations.

Identification of Social Problems and Unmet Agency Needs

The first step in the PPI process is identification of societal challenges or public agency mission-related needs which could be addressed by PPI (Edquist and Hommen, 2000; Edler et al., 2005; Edquist et al., Chapter 1 in this volume). Articulating demand for new products satisfying an unmet need might take a variety of forms. It can be triggered by identification of a grand societal challenge (Edquist and Zabala-Iturriagagoitia, 2012) such as climate warming or changing demographic structure of the population. Upcoming regulatory changes, such as more stringent energy efficiency norms for buildings, create the need to develop new technological solutions to meet those regulatory requirements. It has been noted, however, that there is often a disconnection between grand societal challenges, as identified in the national policy arenas, and local service delivery. National problems are not automatically experienced as equivalent problems at an agency or local government level (Rothwell, 1982).

Opportunities for using innovation procurement can be also opened up by investment needs, both for new systems and assets as well as replacement at the end of the lifecycle (for example, hospital, public transportation fleet, or administrative information technology, IT, system). The need might also be related to problems in operational service production and delivery leading to performance gaps (Feller and Menzel, 1977). In some cases an improvement opportunity is opened up by rapid technical development (for example, digital services). Innovative supplier companies, research organizations or stakeholder groups might identify an opportunity to improve the delivery of public services with new innovative products, services or system solutions, and propose their development and adoption to public authorities.

Public organizations thus need to build routines to identify opportunities for PPI. Related to various channels through which needs identification might take place, these routines need to be integrated in the processes the organization has for anticipating policy changes, setting organizational strategy, planning investments, quality improvement of service delivery, and collaboration with suppliers and stakeholders.

Capability 2: To identify societal challenges, unmet service needs and performance gaps, and to anticipate future investments as appropriate opportunities for effective use of public procurement of innovation.

Market Engagement

When a public agency addresses an unmet need or a performance gap, it will need to scan for available new solutions on the marketplace, and also look for capabilities and ideas from suppliers to develop an innovative solution to meet the need. It will often not be possible to know *ex ante* whether an innovative solution satisfying an unmet need will be found on the marketplace or not. Therefore, the decision to select one of the two principal modes of PPI – developmental or adaptive (Edquist and Hommen, 2000) – can only be done after an initial scanning of markets. Developmental procurement occurs when a government agency places an order for something which does not exist but could be expected to be developed. Adaptive procurement is the purchase of a solution which already exists, but has not been adopted yet by other organizations, or which requires a significant amount of additional development for efficient adoption. Choosing between developmental and adaptive procurement requires good knowledge of the market, assessment of available solutions and understanding of the product life cycle. It is also necessary to resist temptation to support the generation of new technology through public procurement when it is already available elsewhere (Geroski, 1990).

Capability 3: To search for alternative solutions, assess the technical maturity of suggested solutions and their product life cycle, and select between developmental and adaptive procurement according to the available supply on the market.

There is a greater need for interaction with suppliers in PPI than in regular procurement of standard products. Information about procurers' unmet needs must be communicated to potential suppliers. In return, suppliers' knowledge of possible technological solutions needs to be transmitted to procurers (Edler et al., 2005). Appropriate linkage mechanisms with suppliers need to be established (Rothwell, 1984). By communicating its innovation needs to the market, government might encourage firms to invest in product development as a prospect of recovering profits from future sales of the product (Lichtenberg, 1988). By doing so government may avoid, partially or completely, procuring product development from firms, thus placing most of the technology risks with the suppliers. The commitment to purchase an innovative solution in the future (by using functional specifications) might provide enough incentive for firms to initiate innovating to meet government needs.

Technology and innovation studies emphasize the pivotal role of user–producer interaction and interorganizational learning as a basis for innovation (Lundvall, 1985; Rothwell, 1994). Brought into the context of PPI, the learning process between government buyers and private suppliers is constrained by tighter procedural rules for interaction than in private procurement. A specific organizational capacity needs to be built within public buyer organizations for effective market engagement approaches which are applicable within the boundaries of public procurement regulations and state-aid rules. Essential for user–producer interaction is exchange of information regarding future needs of the government and suppliers' technological and organizational capability to respond to these needs (Edquist et al., 2000). Various legitimate market engagement techniques to be used at the pre-procurement phase are available, such as use of reference groups, joint foresight and market sounding methods (OGC, 2009).

Capability 4: To interact with suppliers to communicate agency needs, collect market information and engage with them in order to stimulate their innovation activities responding to agency needs.

Market interaction practices should also be receptive to new ideas and techniques suggested by suppliers (Roessner, 1979). Valuable new opportunities to improve the performance of service delivery process might be opened by innovative solutions from markets. However, partnering a single firm with a promising idea often conflicts with fairness and non-discrimination rules related to public procurement. Yet, to neglect potential innovation simply for the reason that the initiative has come from suppliers, not the procuring organization itself, valuable innovation opportunities can be wasted. Therefore procuring organizations need to create practices to scan, collect and evaluate innovation proposals from suppliers and evaluate them against policy goals. If an organization's capacity to carry out that type of scanning is limited, it can substitute some of that work by being engaged in joint research projects, foresight activities and professional networks.

Capability 5: To process innovation ideas suggested by supplier firms and assess them against policy goals and agency needs.

Rothwell claims that 'it is when government is itself the end user of the product, rather than acting as a broker or middleman, that public procurement is potentially most effective' (Rothwell, 1984, p. 325). Government users should understand their own needs well and formulate requirements accordingly. They can also communicate these requirements directly to potential suppliers and signal long-term commitment to these needs.

There is a need to understand the market power of the government buyers, the structure of the supply side, and knowledge of supplier capabilities in order to reap full benefits from PPI (Mowery and Rosenberg, 1979; Rothwell, 1982; Mowery, 2002). Government can be the lead user for certain industries and act as a launching customer or the first user (Dalpe et al., 1992). In these cases, government has the potential power to structure the market because of its large purchasing volume or the high performance requirements it imposes. However, this is not the case in all industries or market segments. Government may have only a weak or moderate power to influence market structure and direction of innovation through its purchasing activities. Being just another buyer among many others may restrict the possibilities of using PPI effectively.

Capability 6: To understand government's market power to drive innovation and influence the structure of the supply side.

Moreover, the market power of government is often fragmented into distinct departments and agencies with little coordination (Caldwell et al., 2005). While in aggregate terms the government in many policy domains commands a substantial share of demand, its fragmented implementation leaves public procurers with less of an upper hand in relation to markets. Bundling this demand across organizations and departments creates larger incentives for firms to invest in product development targeting a larger and more varied market (Roessner, 1979; Dalpe, 1994). It could also stimulate the emergence of a lead market for a new product or industry (Beise, 2004; Beise and Cleff, 2004). Cross-organizational coordination thus appears as an essential management capability. Good practices, such as common procurement units and procurement networks, are already available in many policy domains and can be used for PPI.

In the same vein, bundling demand across countries within a regional trade area, such as the European single market, creates larger markets for innovative products. This is particularly important for sectoral government agencies, which with their advanced and highly specific needs can be lead buyers for their supplier industries. Bundling the demand at the supranational level opens larger markets for specialized products and services (Timmermans and Zabala-Iturriagagoitia, 2013). The lack of uniformity in procurement codes and other regulations, however, makes aggregation of markets a challenging task (Roessner, 1979). International collaboration projects are usually needed in order to align procurements across borders.

Capability 7: Cross-organizational coordination to bundle demand within the government and international collaboration to create larger markets across countries.

The ultimate innovation policy goal for PPI is that new products and services also diffuse to other government users, private users and export markets. Understanding the preconditions for a diffusion process is an essential function for innovation procurement (Rothwell, 1982; Rolfstam et al., 2011). There is a particular risk that new products and solutions will not spill over to private markets (Tsipouri et al., 2010). Poor diffusion of innovation can undermine higher investments in procurement. A particular challenge is related to complex procurements with a high degree of uniqueness in design (for example, a public building), with tailored features to match specific agency needs and technical interfaces (for example, an administrative IT system), or with a significant service component (for example, a human care service). In these types of instances, replication of innovation to other clients is not commonplace, thus reducing diffusion potential. The ability to analyse and understand the preconditions for diffusion thus appears to be a particular requirement for exploiting PPI to generate extensive economic and social impact. Consideration should be given to both supply- and demand-side characteristics: the replicability and scalability of the product, capacity of the firm to reach out to other markets, and the absorptive capacity of the market targeted for diffusion.

Capability 8: To understand the preconditions under which innovation generated by public procurement could diffuse to other clients, private users and export markets.

Looking at the issue of diffusion from the firm's perspective, it translates into the question of whether firms have sufficient incentives to commercialize and sell products and services to broader markets. This requires a sufficient level of intellectual property to remain with the supplier (Nyiri et al., 2007). Complex contractual arrangements are often needed, particularly in complex procurements, to ensure that the necessary incentives for the companies to invest in innovation are available through intellectual property rights (IPR), at the same time as allowing sufficient user rights on the buyer side to exploit a new product or service. Assigning intellectual property is a specific topic with many highly technical aspects of great relevance for managing PPI.

Capability 9: To assign and manage intellectual property in a way which enables exploitation of innovation by suppliers to scale up business to other clients and markets.

Specification of Functional Requirements

Specifying requirements in functional terms and not dictating particular technical solutions leaves room for suppliers to suggest alternative technological solutions (see Chapter 1 in this volume). Functional requirements appear under a variety of names across sectors: performance-based requirements (Buchanan and Klingner, 2007), outcome-based contracts (Ng et al., 2009) and value-driven procurement (Dumond, 1996). What all these concepts have in common is the focus on functions, outcomes and performance of the procured product, not its technical features or design. The supplier is thus given freedom to suggest alternative solutions and new technologies to generate the expected value to the procuring organization.

Specifying the functions, performance or value that the procured product is expected to deliver is best done by actual users of the prospective product (Geroski, 1990; Dalpe, 1994; Edler et al., 2005). Users have the best knowledge about their real needs and can therefore contribute most to a supplier's innovation. In some cases, particularly when the procuring unit is the actual user of the acquired product, users can be engaged at low additional cost. In other instances, particularly when end users are citizens consuming contracted-out public services, engaging users requires specific efforts and may incur significant extra costs. Designated user engagement practices need to be set in place. Reconciling expectations and needs from a large number of users, especially among a heterogeneous set of users, may require a considerable amount of time and effort (Edler et al., 2005).

Capability 10: To engage users in order to incorporate their needs in functional requirements and increase acceptance of new prospective products among them.

It has been argued that purchasing authorities need to have considerable technical expertise in order to specify functional requirements and evaluate various technical solutions offered by suppliers (Rothwell, 1984; Rolfstam, 2007). It has also been observed that firms have a greater incentive to innovate if government is capable of providing them with technical support (Dalpe, 1994; Palmberg, 2002), and thus engage in a collaborative user–producer innovation process. Technical expertise can be provided by in-house experts, or it can be contracted from external expert service providers such as consultants and research organizations. Some of the required expertise may also become available through collaborative research and development projects. As market-based approaches to public service delivery have gained ground through contracting out and privatization, the technical capacity of the public sector has also become thinner. Government has thus become more reliant on external technical specialists.

Capability 11: To have technical and managerial skills in place needed to develop functional requirements and evaluate alternative solutions proposed by suppliers.

Piloting and Testing Innovations

When a public agency places an order for something which does not yet exist in the market, or intends to purchase a novel product not yet tested by other users, the agency needs to confirm that the product functions as intended and delivers the expected performance. The use of functional requirements in procurement calls for verification of those functionalities, which can be conceptualized as product performance, quality and quantity of outputs, or effectiveness of public services. Testing arrangements are needed to be put in place. They can take advantage of external technical services such as testing service providers and technical research organizations.

Mere technical testing is often not enough to ensure the effectiveness of technology adoption. A change of organizational practices is needed because technology seldom fits the user environment in a seamless fashion (Leonard-Barton, 1988). Innovations transform the organizational structure and practice. At the same time, innovations need to adapt to existing organizational arrangements. The mutual adaptation of innovation and user organization leads to a process of 'inno-fusion' (Fleck, 1993). Many technologies are configurational, requiring adjustment to other surrounding technologies, organizational practices and institutions. Especially when the product is expected to function as part of a larger technical system – such as an information technology system – integration with other organizational practices and other administrative information systems becomes critical.

The adaptation of innovation to an organizational context

takes place most effectively when actual users are engaged in the development process. Organizational change processes should be organized concurrently with product development linked by piloting projects engaging users. The engagement of users can be arranged by setting up testing and experimentation environments specifically designed to allow piloting of new products and solutions with users, and to experiment with product variants (Geroski, 1990).

Against this background, it becomes evident that there is a need to create better links between PPI and R&D. R&D activities may contribute to market analysis, technology development activities, user engagement, development of functional specifications, and developing and exploiting standards. The key issue is to create appropriate integration of R&D and procurement as a sequenced process taking into account the technology maturity. Advanced technology procurement organizations have developed sophisticated approaches to establishing this integration (Mowery, 2002). It has been noted that creating functioning links between procurement and R&D might be easier to establish for procuring organizations with a strong mission orientation and strong capacity in R&D. This is common specifically in the sectors of defence, space and energy, and to a lesser extent in other domains. Linking with the organization's R&D activities highlights the integration of PPI as a demand-side approach with supply-side policies (Dalpe, 1994; Edquist and Hommen, 2000).

Capability 12: To organize piloting and testing activities linking users to enable suppliers to develop better solutions and to verify the performance of new products and services.

Conventional public procurement is based on a transactional approach which keeps suppliers at arm's length from public buyers. PPI calls for a more relational and interactive approach to engaging with market players. A topical discussion in the general procurement literature emphasizes relationship management as opposed to transactional contracting. The complexity of services procured, consisting of bundles of service and product characteristics, and shifting towards contracting performance over an extended lifecycle, is considered to require partnership-based approaches (Schapper et al., 2006; Caldwell and Howard, 2011). Many cases of complex procurements involve public buyers and private suppliers entering into a relationship based on more than contractual arrangements, in order to facilitate innovation over the contract period.

While a relationship-based management approach would be called for from an innovation promotion perspective, it is clear that there is a risk of encouraging favouritism, oligopoly and artificial creation of barriers to new entrants (Caldwell et al., 2005). Utilizing a more relationship-based management approach is a challenging task, as public procurement rules are based on strict procedural rules with regard to communication with suppliers. However, the rules are particularly constraining when a bidding process has been formally initiated. In the pre-commercial phase which precedes the publication of a call for tenders, rules are not as restrictive and allow more freedom in interaction between public buyers and their suppliers.

Capability 13: To build supplier relationships which enable innovation while respecting public procurement norms of fairness and non-discrimination.

Managing Risks

Expected efficiency and effectiveness gains from PPI cannot be guaranteed due to the inherent risks involved with any innovation. Technological risks may lead to non-completion, underperformance or false performance of the procured service or product (Edler and Georghiou, 2007; Tsipouri et al., 2010). Introducing explicit objectives to promote innovation through procurement essentially increases the risks, as outcomes of procurement of innovation cannot be controlled to the same extent as when procuring off-the-shelf products and services. A higher degree of risk might, particularly in the short run, undermine the potential improvement of value for money. The realization of efficiency and effectiveness gains may also extend over a longer period of time than in traditional procurement.

Managing technological risks related to lower than expected performance of the procured service or product is a key function in the management of PPI. There is a range of proposed approaches to managing technology risks (Tsipouri et al., 2010; Edler et al., Chapter 4 in this volume). However, it needs to be acknowledged that the risks can never be fully eliminated since some of the innovation efforts fail due to uncertainty of the outcomes. This risk can be dealt with in three principal ways. First, risks and expected rewards can be balanced by establishing a portfolio of innovation procurement projects at a sector policy or agency level. Even if a single PPI project fails, the other successful ones in the portfolio will level it out and make it easier to justify the investments made. Second, the risk can be shared by establishing co-financing schemes with organizations providing public innovation funding. The mission of these agencies is to encourage investments in innovation by sharing some of the risks through public funding. This leverage can also be used in conjunction with public procurement.

Third, the process can be split into separate pre-commercial and commercial phases where firms bear most of the risks in the pre-commercial phase (Timmermans and Zabala-Iturriagagoitia, 2013). This phase, which possibly includes an R&D contract, does not automatically lead to the commercial contract, but is subject to open bidding. Product development can be supported in the pre-commercial phase by creating linkages to user needs and allowing suppliers to pilot their products in operational service delivery contexts. The commercial phase procurement can be opened up for competing options by using functional specifications, thus reducing the technological risks involved.

Capability 14: To manage technological risks involved with procuring something which does not yet exist or has not been widely tested on the marketplace.

In addition to technological risks, organizational and societal risks may also undermine potential benefits from innovation procurement (Tsipouri et al., 2010). Lack of acceptance from the users or the society of the new or transformed service is another major risk in PPI. Various approaches to engaging users in the development process and testing of new services and products are available to manage these risks. Promoting and managing the acceptance of innovative services and products procured should be part of the management framework for PPI.

Capability 15: To promote acceptance of innovations among users and manage risks of unsuccessful integration with existing technical systems and organizational practices.

Coordination, Communication and Capacity Building

PPI requires effective coordination within administrative functions and units to accommodate their different targets and incentive structures (Edler and Georghiou, 2007; Magro et al., 2014). Both horizontal and vertical coordination is called for. On the interorganizational level, the coordination of various ministries and authorities is needed. In particular, horizontal alignment is needed between sector agencies and innovation policy organizations. Within sectors, the coordination of societal challenges and PPI initiatives benefits from the vertical coordination of ministries, agencies and local government. Particularly, the objective of bundling demand into larger sets of procurement requires effective coordination efforts. In some cases this might also involve centralized government purchasing units.

Within organizational boundaries, horizontal coordination should take place between service delivery units, procurement units, and research and development functions. Domain expertise within service delivery units needs to link effectively with procedural knowledge residing with procurement functions.

The vertical alignment of politically elected policy-makers, general management and operational level is needed. From the policy-makers' perspective, coordination is needed to ensure that the identified societal challenges are translated into appropriately framed PPI projects. From an operational perspective, effective communication efforts are required to obtain the necessary buy-in for the higher level of risk-taking involved with PPI. Elected policy-makers are generally perceived as risk averse and biased towards short-term projects with immediate pay-offs realizing within an electoral cycle (Edquist and Hommen, 2000).

Capability 16: To coordinate and communicate public procurement of innovation vertically and horizontally at both interadministrative and intra-organizational levels.

It has been noted at various points earlier in this chapter that PPI requires a set of organizational capacities and individual skills which are not mainstream professional qualifications in public administration. Capacity building needs to be enhanced in order to strengthen public agencies' readiness to initiate and execute PPI (Rothwell, 1984; Edler et al., 2005). The procuring agency needs to be able to understand and assess technological solutions offered by suppliers. These capacities can be built through both formal and informal learning mechanisms, such as training programmes, curriculum development, peer learning networks, organizational benchmarking and R&D projects.

The technical expertise can be built in-house or sourced from external experts. While a significant amount of expertise can be insourced, a public organization needs a certain amount of capacity in-house.

Capability 17: To initiate systematic capacity building to increase readiness of the public agency to carry out public procurement of innovation.

CONCLUDING REMARKS

The discussion above has identified a set of management functions which need to be established at an organizational level in order to implement PPI effectively. Some of these managerial functions are more clearly capabilities within organizational boundaries, while others have a more distributed nature within government administration and collaboration partnerships.

The functions have been derived from the literature on PPI, innovation research and studies on public procurement. The main argument of this chapter is that these functions are the general requirements for management that need to be established for an effective deployment of PPI as an organizational capacity, rather than as improvised projects on an ad hoc basis. These requirements add to the list of regular procurement skills, such as writing specifications, managing bidding processes, bid evaluation, and contract awarding and management.

How these functions are to be implemented in each public agency depends on multiple local contingencies which warrant context-sensitive adaptation. Making uniform, one-size-fits-all recommendations about appropriate tools would not be justified, as public agencies vary considerably in size, mission, organizational model and level of initial competence. By focusing on the general management functions, I have aimed to highlight not only evident technical competencies, but also some of the key tensions that complicate deployment of PPI.

The list of management functions also provides a shortlist for innovation policy-makers aiming to strengthen the capacity of public administration to execute PPI. The measures can include various initiatives, such as establishing guidelines for PPI, setting up awareness-raising campaigns and training programmes, facilitating demand articulation through collaborative foresight, providing co-financing schemes, supporting bundling of demand within public agencies, or creating linkages between PPI, R&D activities and testing and experimentation environments. Effective collaboration between innovation policy-makers and sector agencies can benefit from correct identification of appropriate opportunities to support building capacity for PPI.

Incorporating an innovation perspective in the procurement practice requires improving and expanding organizational capabilities. In effect, the issue covers a larger set of issues beyond procurement, extending to governance issues, management of renewal in public services, managing market interactions with the suppliers, and coordination with research and development. PPI requires a strategically sophisticated understanding of the use of it as a policy tool. What is needed is a management practice which allows a certain degree of risk-taking when it comes to innovation, while maintaining transparent, non-discriminatory and legally sound procurement practices. Finding effective organizational routines for managing this process will take time.

Finally, the chapter highlights the need to incorporate PPI effectively into the broader framework of public sector renewal. While promotion of private sector innovation among suppliers remains a major rationale for PPI, its implementation will be best accomplished when it delivers distinct value to governments struggling to improve public services, meet new service demands, and balance their budgets amidst financial austerity. Encouraging and exploiting supplier innovation have the potential to both deliver the economic benefits of market innovation and solve the societal challenges and public service delivery bottlenecks. This positioning must also acknowledge that PPI will not be the optimal instrument in all instances. If anything, it needs to be used in combination with other instruments as part of effective policy mixes. Finding the appropriate application space for PPI remains a challenge calling for more practical applications by public agencies and further conceptual work by innovation scholars.

REFERENCES

Arrowsmith, S. (2010). Horizontal policies in public procurement: a taxonomy. *Journal of Public Procurement*, 10(2), 149–186.
Aschhoff, B., Sofka, W. (2009). Innovation on demand – can public procurement drive market success of innovations? *Research Policy*, 38, 1235–1247.
Becker, M., Lazaric, N., Nelson, R., Winter, S. (2005). Applying organizational routines in understanding organizational change. *Industrial and Corporate Change*, 14(5), 775–791.
Beise, M. (2004). Lead markets: country-specific drivers of the global diffusion of innovations. *Research Policy*, 33, 997–1018.
Beise, M., Cleff, T. (2004). Assessing the lead market potential of countries for innovation projects. *Journal of International Management*, 10, 453–477.
Buchanan, N., Klingner, D. (2007). Performance-based contracting: are we following the mandate? *Journal of Public Procurement*, 7(3), 301–332.
Caldwell, N., Howard, M. (2011). *Procuring Complex Performance: Studies of Innovation in Product-Service Management*. Routledge, New York.
Caldwell, N., Walker, H., Harland, C., Knight, L., Zheng, J., Wakeley, T. (2005). Promoting competitive markets: the role of public procurement. *Journal of Purchasing and Supply Management*, 11, 242–251.
Dalpe, R. (1994). Effects of government procurement on industrial innovation. *Technology in Society*, 16(1), 65–83.
Dalpe, R., DeBresson, C., Xiaoping, H. (1992). The public sector as first user of innovations. *Research Policy*, 21, 251–263.
Dumond, E. (1996). Applying value-based management to procurement. *International Journal of Physical Distribution and Logistics*, 26, 5–24.
Edler, J., Georghiou, L. (2007). Public procurement and innovation – resurrecting the demand side. *Research Policy*, 36, 949–963.
Edler, J., Ruhland, S., Hafner, S., Rigby, J., Georghiou, L., Hommen, L., Rolfstam, M., Edquist, C., Tsipouri, L., Papadakou, M. (2005). Innovation and public procurement: review of issues at stake. Study for the European Commission (No ENTR/03/24). Fraunhofer Institute Systems and Innovation Research.
Edquist, C. (2011). Design of innovation policy through diagnostic analysis: identification of systemic problems (or failures). *Industrial and Corporate Change*, 20(6), 1725–1756.
Edquist, C., Hommen, L. (1999). Systems of innovation: theory and policy from the demand side. *Technology in Society*, 21, 63–79.
Edquist, C., Hommen, L. (2000). Government technology procurement and innovation theory. In Edquist, C., Hommen, L., Tsipouri, L. (eds), *Public Technology Procurement and Innovation*. Dordrecht: Springer, pp. 5–70.

Edquist, C., Hommen, L., Tsipouri, L. (eds) (2000). *Public Technology Procurement and Innovation*. Dordrecht: Springer.
Edquist, C., Zabala-Iturriagagoitia, J.M. (2012). Public procurement for innovation as mission-oriented public policy. *Research Policy*, 41(10), 1757–1769.
Eliasson, G. (2010). *Advanced Public Procurement as Industrial Policy: The Aircraft Industry as a Technical University*. London: Springer.
Feldman, M., Pentland, B. (2003). Reconceptualizing organizational routines as a source of flexibility and change. *Administrative Science Quarterly*, 48, 94–118.
Feller, I., Menzel, D. (1977). Diffusion milieus as a focus of research on innovation in the public sector. *Policy Sciences*, 8, 49–68.
Fleck, J. (1993). Configurations: crystallizing contingency. *International Journal of Human Factors in Manufacturing*, 3(1), 15–36.
Geroski, P. (1990). Procurement policy as a tool of industrial policy. *International Review of Applied Economics*, 4(2), 182–198.
Hekkert, M., Suurs, R., Negro, S., Kuhlmann, S., Smits, R. (2007). Functions of innovation systems: a new approach for analysing technological change. *Technological Forecasting and Social Change*, 74, 413–432.
Kalvet, T., Lember, V. (2010). Risk management in public procurement for innovation: the case of Nordic–Baltic Sea cities. *Innovation*, 23(3), 241–262.
Leonard-Barton, D. (1988). Implementation as mutual adaption of technology and organization. *Research Policy*, 17, 251–267.
Lichtenberg, F. (1988). The private R&D investment response to federal design and technical competitions. *American Economic Review*, 78(3), 550–559.
Lundvall, B. 1985. *Product Innovation and User–Producer Interaction*. Aalborg: Aalborg University Press.
Magro, E., Navarro, M., Zabala-Iturriagagoitia, J.M. (2014). Coordination-mix: the hidden face of STI policy. Review of Policy Research, 31(5), 367–389.
Mowery, D. (2002). Defense-related R&D as a model for 'grand challenges' technology policies. *Research Policy*, 41(10), 1703–1715.
Mowery, D., Rosenberg, N. (1979). The influence of market demand upon innovation: a critical review of some recent empirical studies. *Research Policy*, 8, 102–153.
Nelson, R., Winter, S. (1982). *An Evolutionary Theory of Economic Change*. Cambridge: Belknap.
Ng, I., Maull, R., Yip, N. (2009). Outcome-based contracts as a driver for systems thinking and service-dominant logic in service science: evidence from the defence industry. *European Management Journal*, 27, 377–387.
Nyiri, L., Osimo, D., Özcivelek, R., Centeno, C., Cabrera, M. (2007).

Public procurement for the promotion of R&D and innovation in ICT. Institute for Prospective Studies, Joint Research Centre, European Commission. Available at http://ftp.jrc.es/EURdoc/eur22671en.pdf.

OGC (2009). Early market engagement principles and examples of good practice. London: Office of Government Commerce.

Palmberg, C. (2002). Technological systems and competent procurers – the transformation of Nokia and the Finnish telecom industry revisited? *Telecommunications Policy*, 26, 129–148.

Piga, G., Thai, K. (2007). *Advancing Public Procurement: Practices, Innovation, and Knowledge-Sharing*. Boca Raton, FL: PrAcademics Press.

Roessner, J. (1979). The local government market as a stimulus to industrial innovation. *Research Policy*, 8, 340–362.

Rolfstam, M. (2007). Public procurement of innovations: the case of the energy centre in Bracknell, UK. Paper presented in the DRUID-DIME Academy Winter 2007 PhD Conference on Geography, Innovation and Industrial Dynamics.

Rolfstam, M., Phillips, W., Bakker, E. (2011). Public procurement of innovations, diffusion and endogenous institutions. *International Journal of Public Sector Management*, 24(5), 452–468.

Rothwell, R. (1982). Government innovation policy: some past problems and recent trends. *Technological Forecasting and Social Change*, 22, 3–30.

Rothwell, R. (1984). Technology-based small firms and regional innovation potential: the role of public procurement. *Journal of Public Policy*, 4(4), 307–332.

Rothwell, R. (1994). Issues in user–producer relations in the innovation process: the role of government. *International Journal of Technology Management*, 9(5–6), 629–649.

Schapper, P., Veiga Malta, J., Gilbert, D. (2006). An analytical framework for the management and reform of public procurement. *Journal of Public Procurement*, 6(1–2), 1–26.

Timmermans, B., Zabala-Iturriagagoitia, J.M. (2013). Coordinated unbundling: a way to stimulate entrepreneurship through public procurement for innovation. *Science and Public Policy*, 40(5), 674–685.

Tsipouri, L., Edler, J., Uyarra, E., Bodewes, H., Rolfstam, M., Sylvest, J., Kalvet, T., Hargeskog, S., Waterman, D., Banciu, D., Vass, I., Creese, S., Theviessen, P. (2010). Risk management in the procurement of innovation. Concepts and empirical evidence in the European Union. Luxembourg: European Commission.

Uyarra, E., Flanagan, K. (2010). Understanding the innovation impacts of public procurement. *European Planning Studies*, 18(1), 123–143.

Winter, S.G. (2000). The satisficing principle in capability learning. *Strategic Management Journal*, 21, 981–996.

4. Risk management in public procurement of innovation: a conceptualization[1]

Jakob Edler, Max Rolfstam, Lena Tsipouri and Elvira Uyarra

INTRODUCTION: THE IMPORTANCE OF RISK AND RISK MANAGEMENT IN PPI

One of the major challenges of public procurement for innovation (PPI) is risk aversion and the limited risk management practices in the public sector (Edler et al., Chapter 2 in this volume). While this problem has been identified for many years (Edler et al., 2005), there have been no attempts to design an effective risk management framework that can be used to alleviate it. The major reason for this is the high level of complexity when it comes to defining, understanding and operationalizing risk, which are necessary in order to make it manageable in the first place. Risks associated with PPI do not only emanate from the nature of the innovation activity itself, but have a large number of different origins associated with the heterogeneity of the actor landscape in PPI. Moreover, different actors have different risk perceptions, and we often find a mismatch between actor groups benefiting from an innovation (users, suppliers, citizens) and those that bear the consequences of its failure.

Against this background, the purpose of this chapter is to conceptualize risk and risk management in PPI and discuss the value of such a conceptualization for PPI practice and policy-making. Our conceptualization of risk and risk management in PPI refers to the various types of risks that are relevant to the public procurement process, and indicates some governance and managerial challenges these pose for PPI. It is not our goal however to present a normative and prescriptive framework, or even a management manual for how to deal with risk. Our assumption is that if risks can be formalized, assessed and communicated, strategies can be devised for risk management and risk mitigation. By doing so, we hope to demystify one of the barriers to PPI, support policy and procurement practice towards a more rational decision-making in PPI, and increase the capability and inclination in the public sector to buy innovation.

For the purposes of this chapter we focus our attention on procurement situations whereby a public agency places an order for a product or system, which does not exist at the time of the call for tenders but which can be developed within a reasonable time period. The chapter includes technology and addresses non-technological innovation and complex systems, where the technology may be known and proven but not on the scale or level of complexity required for the particular procurement. For a buying organization, any product or service that is new to the organization can be defined as an innovation and can pose all sorts of risks associated with innovation adoption. However, for the purpose of a comprehensive risk management framework, we consider the more complex situation which includes the actual generation of innovation and the risks associated with this on the supply side.

In recent years PPI has become an increasingly attractive policy initiative in the European Union for two concurrent reasons:

1. Case studies in the literature – including in this volume – indicate that PPI can be a very effective tool of innovation policy related to both the creation of lead markets and the dissemination of new technology.
2. Demand-side policies are gaining momentum (OECD, 2011; Izsak and Edler, 2011), as they promise to contribute directly to the grand challenge agenda and represent an attractive way to leverage private research and development (R&D) without adding to the budget deficit in the current austerity climate.

Expectations are high for public procurement to contribute to innovation in Europe, mobilize leading-edge private demand and link innovation activities with a better public service. Public procurement applied as a demand-side innovation policy instrument was the cornerstone of the Lead Market Initiative that was launched for six lead markets.[2] The underlying logic has been described in some detail elsewhere (Wilkinson et al., 2005; Edler and Georghiou, 2007; Georghiou, 2007; Rolfstam, 2013), and this logic is compelling.

Despite the increasing emphasis on PPI in all formal European Union (EU) strategy documents and in new experiences in many member states, the response of public procurers themselves is not enthusiastic. Past policy and academic papers analyse the reasons why all these policy recommendations are not translated into more PPI. One major barrier among them is identifying risks and uncertainty associated with the promises of delivery of goods and services that do not yet exist. For example, it has been argued that 'stamina and sophisticated risk management are needed in order to cope with innovations in public services' (Edler and Georghiou, 2007, p. 960). It has also been argued that risk aversion in society, as well as among public procurers, is a serious obstacle to the adoption of appropriate, innovative, reactive and proactive supply strategies (Cox et al., 2005, p. 1; and similarly Nyiri et al., 2007), and needs to be remedied in order to face the challenge of global competition (Aho et al., 2006).

We acknowledge that public procurers are (and should be) risk averse because they carry the responsibility for spending taxpayers' money. But we argue here that, in the case of PPI, taking calculated and planned risks can lead to increased profits, exports and economic growth, and both private and social returns on investment can increase. Risk management comes at a cost, and while such a cost cannot be exactly measured, procuring organizations need to acknowledge it in their budgets. It is therefore essential to get a better grasp of the nature of risk and the actions for risk management, to make sure costs are reasonable and not prohibitive, and to keep in mind that the cost of no risk management may be a lot higher. The main message is that PPI requires someone to take responsibility for an additional investment in risk management. This investment leads to private and/or social returns if the risk management costs do not exceed the overall benefits of the innovation, and are not higher than the opportunity costs of not innovating at all.

From this it follows that conceptualizing and dealing with the risks associated with PPI can help to assess the risk–reward nexus and lead to more PPI in the future. In addition, following the experiences of private sector managers, an organizational culture which allows risk-taking is in itself an indicator of quality and, ultimately, excellence (Porter, 1997).

In order to structure the debate of risk in PPI the chapter first introduces the general concepts of risk, uncertainty and risk management, and then develops a typology of PPI risks by decomposing the various risks that public procurers face along the procurement cycle when procuring innovation. We go on to discuss the basic functions of risk management, and then bring it all together in a discussion of risk management in PPI. Using the typology developed, we try to illustrate how specific characteristics of the procurement itself, the procurement process and its environment are more likely to be linked to some of the risks identified and not others, and how different situations necessitate different risk management strategies. We conclude with a summary of the main points and an appeal for a next step to develop concrete risk management guidance and practices for PPI.

CONCEPTUALIZING RISK IN PPI

Introducing the Concepts of Risk, Uncertainty and Risk Management

To understand and manage risk in PPI, we first need a clear understanding of what risk actually is. There is a range of risk definitions. According to the *Cambridge Advanced Learner's Dictionary* (2008), risk is defined as 'the possibility of something bad happening'. Risk may also be understood in relation to intention, that is, as 'something happening that may have an impact on the achievement of objectives' (NAO, 2000, p. 1), which suggests that risk should not be perceived exclusively as a potentially negative outcome, but also as something potentially good. This definition includes risk as an opportunity, as well as a threat (NAO, 2000, p. 1; similarly OGC, 2003). Based on the classical work by Knight (1921), one may include the notion of measurement in terms of measureable uncertainty. Thus risk can be defined as measureable uncertainty of outcome, whether positive opportunity or negative impact, whereby the measureable uncertainty is expressed in terms of likelihood.

To clarify the difference between uncertainty and risk, Perminova et al. (2008, p. 77) define uncertainty as 'a context for risks as events having a negative impact on the project's outcomes'. Furthermore, risks are known and possible for managers to deal with, while uncertainty refers to an event or a situation not expected to happen.

A broader definition distinguishes between different sources of risk, as risk results:

> from the direct and indirect adverse consequences of outcomes and events that were not accounted for or that were ill prepared for, and concerns their effects on individuals, firms or society at large. It can result from many reasons either internally induced or (and) occurring externally with their effects felt internally. (Kogan and Tapiero, 2007, p. 378)

From this definition it follows that there is a distinction between risk as the result of failures or misjudgements and the result of (for the organization) uncontrollable events (ibid, p. 378); an important distinction when it comes to risk management.

Another way of distinguishing risk, again highly relevant for risk management and risk perception, especially in our context, is to look at demand and supply as sources of risk. Analysing the situation in the toy industry, Johnson (2001, p.106) observes that, 'Demand for fad-driven products can move from tepid to boiling overnight and then suddenly evaporate as the next hot product sweeps the market'. Concerning supply, 'Supply chains that span the globe and include many emerging countries add currency and political risk that can disrupt supply and change cost structures with little notice' (ibid., p. 106). Sodhi and Lee (2007) add another category of risk which is not specific to demand or supply but relates to a specific context. One important contextual dimension is cultural differences which may lead to misunderstandings and conflicts and thus lower operational efficiency (Sodhi and Lee, 2007). On the supply side, specifically, two forms of contextual risk are distinguished in the literature. Contextual supply risks are those that potentially disrupt or delay operations. The reasons for such risks are manifold, ranging from political instability[3] and volatile labour markets to those that originate in the actual supplier's market context, such as disruptive changes in supply chains (for example, takeovers) and inadequate performance of the supply chain, resulting in the inability of the purchasing organization to meet customer demand, or threats to customer life and safety (Zsidisin et al., 2004, p. 397).

Similar concepts have been used for risks associated with innovation in general. Risks involved in radical innovation have been mapped according to three dimensions: the degree of uncertainty, the degree of controllability and the relative importance (in other words, benefit) (Keizer and Halman, 2007). If the likelihood of a bad result is high, the ability and resources available to influence and control outcomes are small and the potential consequences of failure are dire, a project activity should be labelled 'risky' (Keizer et al., 2002).

For the purpose of this chapter, we refer to risk in PPI as any action or event relating to the process of planning, purchasing, implementation and management of goods, works or services, which has an adverse impact, not only on the delivery of public services but also on the generation and diffusion of innovations.

Against this basic characterization of risk, we can now define risk management as the processes and activities with which to identify and deal with the measureable uncertainty of outcome, be it a positive opportunity or a negative impact. It needs to be understood as a risk–reward management, as any risk is to be assessed not only against the likelihood of its occurrence and the negative effects once they occur, but also against the benefits of the procurement for the various actors. The benefit is not absolute, but determined in relation to the overall targets and context conditions of the PPI and the actors involved.

Risks in Public Procurement of Innovations

In order to understand the specific challenges of risks in PPI, we now need to develop a more nuanced picture of such risks. As defined above, they can have different origins and affect different aspects during the procurement process as well as the innovation cycle.

First, five different types of risk can be defined, all derived from the source or type of phenomena potentially leading to or explaining the adverse effects on outcomes or processes. These five categories are based on Miller and Lessard (2008) and Keizer et al. (2002), but are largely extended through our own deliberations and discussions in the context of the preparation of the European Commission report on Risk Management in the Procurement of Innovation,[4] and complemented with arguments from the literature (Zsidisin and Smith, 2005; Cox et al., 2005):

1. Technological risks are all those risks that lead to non-completion, under-performance or false performance of the

procured good and service. Due to the innovative nature of PPI, risk lies in the technical characteristics of the service or product or in its production, and thus originates from the suppliers' side. Technological risk in PPI involving products in the fluid phase appears to be of particular relevance (Utterback and Abernathy, 1978).

2. Organizational and societal risks. Organizational risks are all those risks in PPI leading to failure or underdelivering for reasons within the procuring organization. Societal risks are those related to a lack of acceptance and uptake by the buyers and users of the new or changed service delivered.
3. Market risks are to be found on the demand and supply sides. The former risks occur when innovations in PPI projects are also intended to spill over to private markets, which are not large or responsive enough, or do not build up quickly enough, to justify capacity investment. The latter risks are potentially disrupted or delayed operations due to political instability and a volatile labour market, potential threats such as the supplying firm stops delivering (exits the market, being acquired by a larger firm not interested in delivery, and so on), or a delayed or poor delivery due to insufficient quality of suppliers.
4. Financial risks in PPI are related to uncertainty in meeting target costs and the ability to secure the funds needed.
5. Turbulence risks are associated with large-scale projects and emerge from a range of unforeseen political or economic events that lead various actors in the whole process to reassess their priorities or change their expectations.

Second, to fully understand the risks in PPI, we can further link those five types to the different stages of PPI, defined by two related cycles, that is:

1. The procurement cycle, as risks, risk perception and risk allocation change in the different stages of the procurement process. It should be noted that the procurement cycle is a generic model of a staged procurement process. In practice, individual procurement projects may vary in detail and may also affect the risk profile for any given project.
2. The innovation cycle, as we face different risks in the different stages of an innovation process, such as generation, adoption and diffusion of the innovation.

Table 4.1 shows the different manifestations of the five risk types in the different stages of the procurement process and the innovation cycle. This table is highly simplified and stylized, but it presents a first picture of the variety of risk sources and the change of sources and types of risks in the innovation and procurement process over time.

While this categorization helps in understanding the nature and sources of risk, it is not as yet linked to the various actors involved. In a next step, we bring in three major actor groups: suppliers, procurers and public policy-makers as well as the end beneficiaries (citizens). Only by understanding how different actors are affected by different types of risks (and the consequences should they materialize), and how they (potentially) contribute to the risk occurring, can we can draw conclusions on the cost–reward sharing of carrying the various risks and develop a basis for risk management. Table 4.2 demonstrates that public bodies and policy-making bear considerable responsibility for risk and failure. In addition, it also shows that suppliers and society more generally contribute to failure. Tackling those failures, however, is an integral part of a risk management strategy as well as policies that support PPI more generally.

CONCEPTUALIZING RISK MANAGEMENT IN PPI

Having conceptualized the nature of risk, we can now turn to the discussion of risk management in PPI. This discussion starts with a basic definition of risk management, followed by reflections on the particular management issues associated with PPI.

Basic Functions and Principles of Risk Management

Risk management increases the likelihood of project success and reduces the cost of failure. Studies of new product (Keizer et al., 2005) and software development projects (Bannerman, 2007) have shown that the chances of success increase when risks are managed, including the monitoring and application of reflective learning and sense-making throughout the project. PPI would therefore also benefit from the application of sound risk management principles, although this may not necessarily mean that a formal structure is set up (Chapman and Ward, 2004).

Risk management also reduces innovation costs. For

Table 4.1 The risk map in PPI

	Sources of risk					
Stages in procurement cycle	Organizational institutional/ societal	Financial	Market	Technological	Turbulence	Stages in the innovation cycle
Planning and preparation	Definition risk	Financial planning risk	Supplier market risk	Technical risk		R&D stage
Notification and prequalification	Failure to define needs and communicate to market	Innovation far beyond initial budget	Not enough capable bidders	Solution not feasible or suboptimal		
	Legal/ regulatory	Financial market risk	Supply chain risk	Contract design/ award/ evaluation proc. not adequate for technology		
Tendering Evaluation Contract award	Changes in regulations, misalignment with and procurement objectives	Failure to secure funding	Suppliers taking hidden risks Supply chain deficient		Unforeseen events mainly associated	

Table 4.1 (continued)

	Sources of risk					
	Adaptation risks		Market spillover	Lack of complementarities with networks/ standards	with large scale projects	Adoption by public client
Contract management	Internal integration /external acceptance	Cost monitoring	No spillover to private markets			Diffusion in public realm
				High cost of upgrade and maintenance		
Evaluation		Poor cost controlling, and choice of payment modalities				Diffusion in private markets
	Policy spillover		Market competition risk			Maintenance and upgrading
	No adoption/use by other services/policies		Dependency on few suppliers/ competition distortion	Technological lock-in		New cycle
Risk evolving mainly from procurement				Risks evolving mainly from innovation		

Table 4.2 Consequences and causes of failure along the procurement cycle for the three main actor groups

Innovation risk along the innovation cycle	Producer supply chain	Public user, policy-makers	End beneficiary (the citizen)
R&D fails to deliver	*No delivery, no revenue*	*Delay in service provision, additional time lag, costs*	*No innovative product or service*
	The basic design of R&D was erroneous	Poor contractor involvement, short-term focus	
First adoption by public client failed or delayed	*No closure of contract, no demonstrator*	*No new, improved service, sunk costs*	*Disruption in service*
	Failure to communicate needs to client and to recognize lack of absorptive capacity	Unexpected failure to adapt, internal resistance, switching costs too high, lack of complementarity, leadership	
Diffusion in public realm smaller or non-existent	*High or prohibitive adaptation costs*	*Poor adoption*	*No innovative services in related areas*
	Preferences of other customers not as uniform as expected	Risk aversion, change aversion, competence gap, switching costs, lack of complementarity, lack of leadership	Absorptive capacity/acceptance of beneficiaries insufficient
Spill over in private markets is not realized as expected	*Lack of revenues needed and expected*	*Policy goal (market creation, societal goal) not achieved*	*Society does not benefit from innovation*

Table 4.2 (continued)

Innovation risk along the innovation cycle	Producer supply chain	Public user, policy-makers	End beneficiary (the citizen)
	Failure to identify market opportunities	Failure to anticipate private demand	Lack of sophisticated consumers, prohibitive prices, or counterproductive standards
Subsequent maintenance and updating costly and counter-productive (future lock-in)	*Burden of after-sale maintenance exceeds expectations, higher costs for readjustment of users*	*Runaway costs*	*Discontinuation of innovation, lack of upgrade of innovation*
	Calculations in bid too optimistic	Poor life-cycle costing or commissioning conditions; disruptions because of poor supply-chain management and lack of whole-life costing	
New cycle	*Disadvantage for newcomers and poor future planning*	*No adjusting of policy*	*Readjustments / learning costs, poor satisfaction of changing needs*
	Poor signals as to long-term needs due to lack of strategic prioritization and assessment of needs in policy priorities; exacerbated through over-reliance on public contracts	Technological or organizational lock-in	Poor articulation of future needs

Note: * italic letters depict consequence, Roman letters depict cause.

construction projects, Bauld and McGuiness (2008) have found that public customers in general pay more than private customers, and that this is related to the different use of risk management practices in the public and private spheres. Simply stated, they suggest that the price is affected by the risk assumed by the supplier as follows:

Price to the customer = Supplier's cost of supply +
Risk assumed by the supplier + Profit

From this relationship it follows that the higher the risk public procurers assign to the supplier, the higher the price. Unrealistic allocation of risk built into a contract offered by a public procurer may lead to hidden costs as firms, which know what they are able to deliver (asymmetric information), may choose not to participate in the tender at all (Bauld and McGuiness, 2008). In this sense, risk management in PPI becomes something that goes beyond the boundaries of the procuring public agency. In relation to this, a point has been made that there is also a need for suppliers, especially those which traditionally have not supplied to the public sector, to learn and understand the unique environment of public procurement (Davidson and Moser, 2008).

Against this background, and on the basis of the above typology of risk, there are three major tasks associated with risk management in the context of PPI:[5]

1. Define and assess risks and rewards for all partners involved at the various stages of the procurement process, including (see also AIRMIC et al., 2002; NAO, 2000): the nature or kind of risk (which may change during the various procurement stages), the causes and sources of risk, the likelihood of risks occurring and the potential consequences (additional costs, reduced benefits).
2. Once risks are defined one can, for each type of risk, take action to avoid or reduce the likelihood of risk materializing – that is, the likelihood of an adverse project outcome (Bannerman, 2007) – and allocate responsibilities to take action to reduce this likelihood.
3. Define action plans to mitigate the potential consequences in case a risk occurs, and determine who bears the cost of mitigation and who carries the burden of reduced benefits (contingency plan).

The conceptualization made above points at types of risks, which may occur when innovative products or services are procured. But not all of them occur or are equally relevant. On the contrary: risks, their origin, identification of the best way to decrease the likelihood of risk or mitigate its impact and, finally, the ways to share the reduction and mitigation, are specific to each case. Keeping in mind the uniqueness of each project, we nevertheless argue that there are certain characteristics of the innovation, and of the procurement process itself, which influence the nature and consequences of risks. These are discussed in more detail in the next section.

Risk Management Derived from Project and Process Characteristics

As mentioned above, the specific innovation features of the PPI project and the characteristics of the public procurement processes have important implications for our three main tasks of managing risk in PPI, namely: (1) the type and origin of risk, the ways to reduce the likelihood of the risk occurring; (2) the ways to mitigate the consequences when the risk materializes; and (3) the ways to share the costs among the actors involved. In the following, we outline how innovation characteristics and procurement process characteristics influence risks, and what this then means for risk management and mitigation. This is to be understood as an illustration, a first step towards a broader risk management concept in PPI.

Risks and remedies related to key characteristics of the innovation

In terms of innovation characteristics, we can distinguish between the maturity of the technology, the time anticipated for the procured goods or services to reach the market, their complexity and the budget required for their development.

PPIs (that is, products or services) with low technological maturity are characterized by high technological, societal and demand-side market risks. R&D is risky and the innovation process chosen by the bidder may fail, not deliver a satisfactory solution or deliver a solution that is rapidly overtaken in the market. At the same time, even if an appropriate technical solution is found and delivered, this may prove too costly or too complex for the users to adopt (as in many e-government projects; see Chapter 8 by Caloghirou et al. in this volume). The origin of the technical risk when technological maturity is low lies mainly with the supplier (failed design, lack of expertise), but the consequences will be carried by the procurer and the beneficiaries (if other than the procurer) as well.

Reducing the likelihood of a technological risk occurring may involve a degree of market intelligence by the procurer at early stages of procurement (planning and preparation, pre-qualification procedures), and the use of multiple contracts to select among the best solutions. The likelihood of inadequate R&D design can also be reduced if the contractor carries out complementary R&D in parallel, with internal funding, through other public R&D support or through R&D collaborations in related technologies. Early involvement of the users in the PPI design and extensive awareness-raising and training can reduce the likelihood of societal risks. Market demand risks can only be reduced with appropriate assessments of latent market demand. This may, however, be difficult to do in projects with limited technological maturity.

Mitigation strategies for technological risks may be limited to two main forms. First, additional (renewed) R&D can be a contingency to deliver the desired outcome, even if at a later stage and higher costs. Second, provisions for sharing the costs related to the consequences, should the technological risk occur, ought to be made early on. However, this raises issues about how to estimate such costs. For instance, clauses for additional R&D may entail shifting the cost of mid-design to the supplier.

The scale and complexity of a PPI are other characteristics to be taken into consideration, mainly in terms of market supply and financial risks, but also organizational risks to some extent. While scale and complexity are not necessarily synonymous, they are usually closely linked and hence treated together here. Large recycling plants, residential ecosystems and alternative energy plants are examples of this type of PPI. Projects on a large scale and with high complexity can typically go wrong because of underestimating the barriers, time and cost to reconfigure existing knowledge and to understand the challenges associated with large scale. At the same time, large scale usually needs a high degree of pre-financing and thus access to the financial markets. The budgets for PPI projects can range from a few thousand euros (most often encountered in the provision of specialized software) to multi-million euro projects, when either very innovative technologies are developed (in particular when disruptive technologies are tested, as in the defence sector), or for very large-scale infrastructure (such as long bridges in seismic areas, or lighting entire cities). Financial risks are high for costly projects since, in addition to the usual risks of not meeting target costs, the risk of not securing the funds needed increases as well. Commercial banks are unlikely to be willing to fund risky projects unless they add a considerable premium (and occasionally not even then).

The origin of risk in large-scale projects would start with the procurer. Deep technical knowledge is needed to assess what the change of scale entails. This may require state-of-the-art knowledge and hence cooperation with external advisers to reduce the likelihood of organizational risks, implying additional costs. At the same time, once the tender is launched, suppliers may overestimate their own capabilities to scale up or to access external funding. Supply-side market risks can be reduced if suppliers get expert advice on their side, while financial risks can be reduced if the procurer prepares the ground by discussions with financial institutions. If large societal benefits are expected, public sources such as the European Investment Bank at the European level, or dedicated public financial institutions (such as the German KfW) at the national level, can be approached.

If the supplier fails to deliver (or underdelivers), then clauses to hand over to other interested contractors can mitigate the impact and allow the project to continue after the failure. The mitigation of financial risks can be handled with negotiations with alternative financial institutions at higher costs. Sharing the cost is again a matter of calculation and agreement among the procurer, the supplier, the funding organizations and external public organizations. The financial premium for the complexity of the project, if not accepted by commercial banks, can be carried by private non-profit development banks and by the public budget involved directly in the benefit of the PPI (for example an environmental agency if dealing with alternative energy projects, local authorities if dealing with recycling, and so on).

Finally, time to market is also a parameter that determines risks. The time horizon for a PPI ranges from a few months to many years. Determinants of the time from launching the procedure until the final delivery are technological maturity, the scale and the complexity of the procurement, as well as financial constraint of the procuring body or political processes around the desirability and direction of the innovation, and so on. The increase in risk associated with a longer duration of a PPI has to do with a possible change of preferences of users in public organizations (organizational risk) and of citizens (societal risks), an increase of financing costs (financial risk) and the increased likelihood of turbulence uncertainties.

The likelihood of organizational risk may be reduced by further developing the capabilities of the procuring organization. It is important to focus on keeping long-running PPI well managed in times of reorganizations and administrative changes. Further, the use of multidisciplinary teams, at both the tender and the project delivery stages, and engaging beneficiaries, are all measures to minimize societal risks.

Mitigating the consequences and sharing the cost of organizational risks in long-running projects is difficult. Mitigation may require a new organization taking over, or the involvement of new contractors, either in cooperation with the original supplier or in a new contract. In relation to cost-sharing of organizational risk, whereby the public organization is responsible for inadequacies or restructurings that affect the PPI process, the mitigation investment (hiring competent people or attributing new responsibilities) will have to be carried by the public sector. As for the cost of societal risks in long-running projects, suppliers should contribute as they would be in a position to identify the potential problems early on, adjust the initial proposal and charge the difference to their price. Finally, turbulence cost over longer periods can only be tackled with insurance and should be shared between supplier(s) and procurer(s), as neither of them is in a position to foresee, let alone quantify, it.

Risks and remedies related to the procurement process

In addition to the characteristics of the innovation, the procurement process and context (goals and institutional set-up) are important in determining the likelihood of risks occurring, attributing and managing risks. These are described below.

Firstly, the final purpose of the procurement plays a crucial role. We can differentiate PPI activities that are mainly undertaken to trigger broader demand in the private market (that is, catalytic) from procurement mainly to use the innovation in the buying organization. This catalytic procurement is an extreme case of market creation through PPI, with significant implications for risk management.

In order to create private markets for an innovation, the procuring organization needs good insight into the behavioural characteristics of the beneficiaries or potential adopters of the innovation. Market risks are higher in contexts where the business sector and individual consumers show less inclination towards innovation. As the overall benefit of the procurement lies with this diffusion to other purchasers and users, the societal risk increases considerably.

This leads to a specific set of risk management consequences. The likelihood of insufficient private demand can be reduced by demonstrating the benefits of the innovation (demonstration projects, acting as key reference, and so on), and bundling procurement bids to bring down costs and thus help to activate private buyers. Further means to reduce the likelihood of risk occurring are intense cooperation and coordination between the final users and the procuring body throughout the PPI, training and awareness-raising campaigns, and education measures before the delivery of the PPI.

To mitigate the consequences of societal risk is to find ways to improve and use the delivered PPI *ex post*, once the failure is identified. This means to try to increase the acceptance rates of the beneficiary through awareness-raising and training including feedback loops to redesign the deliverable. Further, trusted committees, composed of experts in different disciplines (technical, financial, social), and from different stakeholder groups for monitoring and early warning, give an opportunity for flexible corrective action. Sharing the cost in this case is between the end user (service provider) and the procuring organizations, potentially in combination with other instruments (beyond the procurement itself), for example, research grants, venture capital and special research guarantees for lead markets. Coordination costs will be carried out for each one of them internally, whereas training and redesign would be attributed to the end user.

The institutional set-up of the buying organization and the national framework plays a very important role when it comes to the generation of risk. It is possible, or even probable, that PPIs with exactly the same innovation and need characteristics will present different risks when performed by organizations with different degrees of technological knowledge and experience in PPI, and in national environments that differ in relation to the propensity to support business R&D and innovation. There are country differences as regards the systematic use and support of PPI. The USA and France have pursued mission-oriented research, the Nordic countries were important examples of radical innovations in the past (ABB with electric power, Ericson and Nokia in mobiles telephony), while the UK, the Netherlands and Germany are increasingly experimenting with pre-commercial variations of PPI and more EU member states are sensitized in this direction (Izsak and Edler, 2011). Further, experiences differ within agencies and national innovation policies, with a range of countries in which experience is limited and institutional risks remain high, associated for instance with a mismatch with the regulatory framework, inadequate specifications or specifications misaligned with the procurement objectives, and miscommunication with other national policies.

Consequently, organizations and systems with a less sophisticated procurement and innovation culture are likely to have higher risks across the board of the spectrum, and might have an inclination for additional risks. For example, inexperienced or ill-trained procurers will need expert advice on the tendering process and coordination with R&D policy support, which may be unwilling to cooperate. The banking system may also be unwilling to experiment with new processes. In those circumstances, reducing the likelihood of the risk occurring may be best addressed by cooperating with experienced agencies and creating early alliances between different national support schemes. A potential means to mitigate the consequences of risk occurring is to accumulate experience by starting with smaller, lower-risk projects. Finally, cost in this case is a matter of national innovation and public procurement policy; neither the procurer nor the supplier is likely to be willing to share a cost deriving from the need to educate procurers and decision-makers in public organizations and national policy-makers.

CONCLUSIONS

This chapter intends to demonstrate that the risk that worries procuring agencies, and that stands in the way of a roll-out of PPI, can be conceptualized and better understood, and eventually managed. Risk is a significant hurdle for PPI, but it is identifiable and measurable, even if by approximation only. Public procurers are aware of risks and are, as a rule, risk averse. They have a lot to lose and little to gain if things go wrong. Hence, their usual reaction is to intuitively avoid PPI or limit it to incremental innovations. Most pioneering examples of PPI were implemented because responsibility was taken at the political level, and maybe the most important lesson from this would be that all risk management only works if backed by the highest decision-makers in organizations.

However, as time goes by and PPI gains momentum, more systematic ways to deal with risk are needed. This chapter contributes to this need by suggesting a typology of risks associated with PPI and investigating how to deal with them. To make risks manageable, we have to understand their differentiated nature defined in a multidimensional space. The first and most obvious dimension is the nature and origin of the risk, related to which we have distinguished technological, organizational and societal, market-related, financial and turbulence risks. To understand the variation within those types, we have further differentiated manifestations of risks along the phases of the procurement cycle, the nature of the innovation, the stages of the innovation cycle and, last but not least, the consequences and the causes of failure for the three most important actor groups involved: producers, buying organizations and citizens.

Against this background, we have decomposed risk management into its basic functions, namely: identify risks and their origin, reduce the likelihood that they manifest themselves, anticipate ways to mitigate the consequences if risks materialize and, finally, allocate risks and their costs. As risks are contextual, we have differentiated further according to the characteristics of the innovation to be procured, the process with which it is implemented, and the institutional and market environment in which it takes place.

The combination of these multiple dimensions finally enable us to start discussing in any meaningful sense the meaning of risk management in concrete situations, linking different categories of characteristics together with the types of risk and the risk management functions. This can then be used to see which risks are more likely to occur in which situations, how to reduce them or mitigate their consequences, and how the cost can be shared.

Our conceptualization will hopefully be the start of a risk management discussion that is tailored towards the specific complexities in the process of procuring innovation in a public organization. It must be part of all stages and interactions, risk occurrence, avoidance and mitigation strategies, and sharing of cost must be built into the process, and the more openly and transparently this is done, the more likely, not less likely, will innovation be. Once risk is defined and managed in this way, the inclination to ask for, offer, generate, buy and adopt an innovation will increase. However, as our initial discussion has shown, there is still a long way to go before risks are understood and managed properly. If public procurement is supposed to be a tool for innovation policy, then surely this must be at the top of the agenda in capability-building for PPI.

NOTES

1. This chapter draws from a report prepared for the European Commission 'Risk management in the procurement of innovation: concepts and empirical evidence in the European Union' prepared by an Expert Group composed by the authors of this chapters and experts preparing case studies: Hanneke Bodewes, Janne Sylvest, Tarmo Kalvet, Sven-Eric Hargeskog, Ditmar Waterman, Doina Banciu, Illona Vass, Sue Creese and Peter Thevissen. We thank all participants of the group for valuable discussions on the concept. http://ec.europa.eu/invest-in-research/pdf/download_en/risk_management.pdf.
2. Sustainable construction, e-health, protective textiles, bio-based products, recycling and renewable energy.
3. Whereby we have to distinguish between those political events that actors perceive to have some likelihood, and those that represent uncertainties, where no likelihood can be attributed as the event is outside the scope of the actor's imagination.
4. http://ec.europa.eu/invest-in-research/pdf/download_en/risk_management.pdf.
5. This list should not be confused with the procurement cycle; it simply illustrates the major issues when it comes to risk in PPI. These tasks are in line with the broader list of tasks defined by the National Audit Office in the UK (NAO, 2000) for risk in public departments more generally, reduced to those that are most relevant for risks in PPI.

REFERENCES

Aho, E., Cornu, J., Georghiou, L. and Subira, A. (2006). Creating an innovative Europe. Report of the Independent Expert Group on R&D and Innovation appointed following the Hampton Court Summit. Luke Georghiou, Rapporteur. EUR 22005. January.

AIRMIC, ALARM and IRM (2002). A risk management standard. London.
Bannerman, P.L. (2007). Risk and risk management in software projects: a reassessment. *Journal of Systems and Software*, archive 81(12), 2118–2133.
Bauld, S. and McGuiness, K. (2008). Yours, mine, ours: offloading all risk risks raising the price. *Summit*, 11(1).
Cambridge Advanced Learner's Dictionary (2008). Available at http://dictionary.cambridge.org/dictionary/british/.
Chapman, C. and Ward, S. (2004). Why risk efficiency is a key aspect of best practice projects. *International Journal of Project Management*, 22, 619–632.
Cox, A., Chicksand, D. and Ireland, P. (2005). Overcoming demand management problems: the scope for improving reactive and proactive supply management in the UK health service. *Journal of Public Procurement*, 5(1), 1–22.
Davidson, S.G. and Moser, S.J. (2008). Rules, regulations, and risk-government vs. commercial contracting. *Contract Management*, 48(4), 34–38.
Edler, J. and Georghiou, L. (2007). Public procurement and innovation – resurrecting the demand side. *Research Policy*, 36(9), 949–963.
Edler, J., Rufland, S., Hafner, S., Rigby, J., Georghiou, G., Hommen, L., Rolfstam, M., Edquist, C., Tsipouri, L. and Papadakou, M. (2005). Innovation and public procurement: review of issues at stake. Study for the European Commission (ENTR/03/24). Fraunhofer: Fraunhofer Institute Systems and Innovation Research, Karlsruhe.
Georghiou, L. (2007). Demanding innovation – lead markets, public procurement and innovation. NESTA Provocation 02: February, London: NESTA.
Izsak, K. and Edler, J. (2011). Trends and challenges in demand-side innovation policies in Europe. Thematic Report 2011 under specific contract for the integration of INNO Policy TrendChart with ERAWATCH (2011–2012). Brussels.
Johnson, M.E. (2001). Learning from toys: lessons in managing supply chain risk from the toy industry. *California Management Review*, 43(3), 106–124.
Keizer, J.A. and Halman, J.I.M. (2007). Diagnosing risk in radical innovation projects. *Research Technology Management*, 50(5), 30–36.
Keizer, J.A., Halman, J.I.M. and Song, M. (2002). From experience: applying the risk diagnosing methodology. *Journal of Product Innovation Management*, 19, 213–232.
Keizer, J.A., Vos, J-P. and Halman, J.I.M. (2005). Risks in new product development: devising a reference tool. *R&D Management*, 35(3), 297–309.
Knight, F. (1921). *Risk, Uncertainty and Profit*. New York: AM Kelley.

Kogan, K. and Tapiero, C.S. (2007). *Supply Chain Games: Operations Management and Risk Valuation*. International Series in Operations Research and Management Science, 113, XII. London: Springer.
Miller, R. and Lessard, D. (2008). Evolving strategy: risk management and the shaping of mega-projects. In Priemus, H., Flyvberg, B. and van Wee, B. (eds), *Decision-Making in Mega-Projects: Cost–Benefit Analysis, Planning and Innovation*, Cheltenham, UK and Northampton, MA, USA: Edward Elgar Publishing, pp. 145–172.
NAO (2000). Supporting innovation: managing risk in government departments. Report by the Comptroller and Auditor General. National Audit Office.
Nyiri, L., Osimo, D., Özcivelek, R., Centeno, C. and Cabrera, M. (2007). Public procurement for the promotion of R&D and innovation in ICT. Institute for Prospective Technological Studies, European Commission. EUR 22671 EN.
OECD (2011). *Demand Side Innovation Policy*. Paris.
Office of Government Commerce (OGC) (2003). Capturing innovation: nurturing suppliers' ideas in the public sector. London.
Perminova, O., Gustafsson, M. and Wikström, K. (2008). Defining uncertainty in projects – a new perspective. *International Journal of Project Management*, 26, 73–79.
Porter, A.M. (1997). In some companies quality culture is tangible. *Purchasing*, 16 January, 122(1), 51–52.
Rolfstam, M. (2013). *Public Procurement and Innovation: The Role of Institutions*. Cheltenham, UK and Northampton, MA, USA: Edward Elgar.
Sodhi, M.S. and Lee, S. (2007). An analysis of sources of risk in the consumer electronics industry. *Journal of Operational Research Society*, 58, 1430–1439.
Utterback, J.M. and Abernathy, W.J. (1978). Patterns of industrial innovation. *Technology Review*, 80(7), 40–47.
Wilkinson, R., Georghiou, L., Cave, J., Bosch C., Caloghirou, Y., Corvers, S., Dalpé, R., Edler, J., Hornbanger, K., Mabile, M., Montejo, M.J., Nilsson, H., O'Leary, R., Piga, G., Tronslin, P. and Ward, E. (2005). Procurement for research and innovation. Report of an Expert Group on measures and actions to assist in the development of procurement practices favourable to private investment in R&D and innovation. Brussels.
Zsidisin, G.A., Ellram, L.M., Carter, J.R. and Cavito, J.L. (2004). An analysis of supply risk assessment techniques. *International Journal of Physical Distribution and Logistics Management*, 34(5), 397–413.
Zsidisin, G.A. and Smith, M.E. (2005). Managing supply risk with early supplier involvement: a case study and research propositions. *Journal of Supply Chain Management*, 41, 44–57.

5. Forward commitment procurement and its effect on perceived risks in PPI projects

Hendrik van Meerveld, Joram Nauta and Gaynor Whyles

INTRODUCTION

This chapter is based on research (Van Meerveld, 2012; Van Meerveld et al., 2012) and the practical experiences of the second and third authors of this chapter in facilitating forward commitment procurement projects conducted as part of the Department for Business, Innovation and Skills (BIS) Innovation for Sustainability Programme and the Low Carbon Building (LCB) Healthcare Network. Although the application of forward commitment procurement (FCP) is limited to a few pilot projects, the outcomes demonstrated are most promising in that they have delivered innovative solutions into the market. Two of the case studies have led to the development of new and innovative solutions and one is progressing towards early adoption of an innovative solution.

The chapter applies a theoretical framework in order to examine in more detail how the FCP methodology works, and to provide insights into how the process could be improved and replicated. Its focus is on the role of the market engagement phase of the FCP process. The chapter adopts a case study approach following Yin (2009), and is aimed at answering the following question: What are the effects of adopting the forward commitment procurement framework on the risks perceived by public sector customers and suppliers in projects concerning public procurement for innovation?

Procurement involves transactions which involve agencies in the form of customers and suppliers. Therefore, agency theory and transaction cost economics are used to develop the theoretical framework. In order to analyse and understand how risks can be managed by applying the FCP method, the study uses these theories to assess the impact of FCP on risks in terms of uncertainty and transaction bounded investments.

The chapter provides three case studies related to the procurement of innovative solutions; two in the UK and one in the Netherlands. From the study it is concluded that the FCP method helps to manage perceived risks in projects aimed at procuring innovative solutions, but its application, like any risk managing process and any innovation process, requires investments on the part of the customer and suppliers.

BACKGROUND

The world is rapidly changing and simultaneously facing a number of grand challenges such as climate change, energy and water supply, public health, ageing societies, financial crises and changes in the world economy (Lund Declaration, 2009). As agents for the social good, public sector organizations are therefore facing considerable operational, financial and efficiency challenges, and at the same time need to become more environmentally sustainable. In order to meet these challenges, public sector organizations will need new, better goods and services. However, these goods and services are not always available in the market, or may not be suitable, effective, or are currently unaffordable. They therefore need suppliers to innovate to deliver what is required.

Over the last decade there has been increasing interest in the use of procurement sector budgets to drive innovation through encouraging and enabling the procurement of innovative goods and services by the public sector (termed public procurement for innovation). Public procurement for innovation (PPI) can be defined as a public sector body procuring goods or services that currently are not available, but can be developed in a reasonable amount of time (Edquist and Hommen, 2000).

According to Edler et al. (2006), risk is an issue of extreme importance in procuring innovative goods or services. Risk is inherent in buying something innovative (European Commission, 2007), and there is a risk of failure. Failure may be total if a supplier is unable to deliver; or partial if performance falls below expectations, or delivery is late (European Commission, 2007).

Failure can also come from practical difficulties in applying new solutions and integrating them within the organization (European Commission, 2007).

PPI presents risks to both the public sector customer and their suppliers that would not be encountered if they were to buy/supply existing and proven solutions. Therefore if public procurement is to be used to drive innovation we need to understand and find ways to overcome these risks. Edler et al. (2006) conclude that risks faced by public sector organizations are different from the risks faced by their suppliers.

Customer Risks in PPI

In essence, the risks that public sector customers face in buying an innovative solution are that: they do not know the outcomes upfront; the new solution might not be delivered on time, might not work or might cost more; and there may be no incentive among suppliers to introduce a new solution to the market.

Several features of the way public sector organizations operate and manage procurement have an impact on the (perceived) risks. Winch (2002) discusses public sector bureaucracies that encourage risk aversion. Public sector organizations are often regarded as risk averse, and the culture of public sector organizations is considered unsupportive or unrewarding of innovation. Where the public agencies are risk averse, this may create barriers to PPI. Risk averse behaviour may be encouraged by:

- Failure to assess the impact of not innovating. In our experience, public sector organizations do not routinely include, in their assessment of risk, the risks that will arise from not innovating. These risks are considerable, given the huge and increasing societal challenges the public sector faces, and the fact that 'business as usual' will result in social costs that are unacceptable.
- Lack of clarity on unmet needs and future needs. Our experience suggests that public sector organizations often find it difficult to accurately determine their needs and define them as outcome-based requirements (for example, functional requirements). Specifying needs as outcome-based requirements is necessary to allow for innovative solutions, but results in uncertainty about what solutions may be developed.

- Information asymmetry. Public sector organizations are often not aware, or not fully aware, of the product and service innovation the market offers them – or could offer them (Edler and Georghiou, 2007). When public organizations do not notice these opportunities it is self-evident that they will formulate their specifications in such a way that currently available goods and services will suffice.
- Lack of know-how in buying innovative goods and services. Public sector organizations experience difficulties in using procurement procedures (for example, tender procedures and contracts, including selection and award criteria) to support innovation. A study undertaken by the Environmental Innovations Advisory Group (EIAG) shows that 66 out of 100 firms surveyed reported that procurement procedures lock public sector purchasers into conventional specifications and do not encourage scaling up to commercial viability of more radical options (DTI, 2006). When there are high levels of uncertainty it becomes more difficult to design the procurement process (Edler et al., 2006). Innovation is hindered if suppliers are unable to participate in tenders (Myoken, 2010; DTI, 2006), or inappropriate award criteria are applied (Edler and Georghiou, 2007; European Commission, 2007).

Not only risks perceived by customers can create barriers to PPI. Suppliers also perceive risks in PPI projects, and these risks may also create barriers.

Supplier Risks in PPI

Innovative goods and services require development (Hommen and Rolfstam, 2009). The costs for the development of new goods and services represent an investment risk. Alternatively, this could also be seen as an opportunity cost. The magnitude of the investment risk in a PPI project depends on the degree of innovation (in other words, radical, incremental) that is required (Edler et al., 2006). In a time of severe economic, social and environmental crisis, entrepreneurs, investors and businesses are particularly exposed when taking investment risks (European Commission, 2011).

Investment decisions are made on the basis of attractive business plans (DTI, 2006), and it therefore follows that, in order to participate in a PPI project, suppliers need to be able to make an assessment of the risks in relation to the potential commercial rewards involved in a PPI project. To do this they need information from the customer. For example, suppliers need to have an accurate understanding of customers' needs and their commitment to buying a solution that meets their needs. However, suppliers of potential new goods and services often lack the knowledge of what customers might want in the future (Edler and Georghiou, 2007), and the lack of clear specification (or perhaps, more accurately, clear information) is a source of uncertainty (Ward and Chapman, 2003).

The underlying risk that suppliers face in developing innovative solutions for the public sector is: having invested in delivering a solution, will the customer buy it (even if it meets their requirements)? If suppliers are to engage in PPI projects, they need customers to tell them accurately what they need, convince them that their need is genuine, and that they are credible customers.

The time of maximum risk for a supplier is not during the research and development (R&D) phase (see Figure 5.1), which is relatively inexpensive and coincides with availability of government funding. Rather, the real risks occur when much larger amounts of money and time are spent on demonstrations and scaling up, before commercial sales prove that the market will buy the product. State aid rules mean that this is a phase when grants are not readily available, while public procurement practice means that very rarely will there be any firm indication that commercial sales will result once the product is available (DTI, 2006). This is the underlying supplier risk that FCP aims to address. The underlying social risk is that without innovation the cost of social needs will become unaffordable, and the agents that society employs to deliver social needs should clearly account for this that is, the cost of innovation needs to be set against the cost of not innovating.

FORWARD COMMITMENT PROCUREMENT

Background

FCP was conceived and the methodology first developed by the Environmental Innovations Advisory Group (EIAG) established

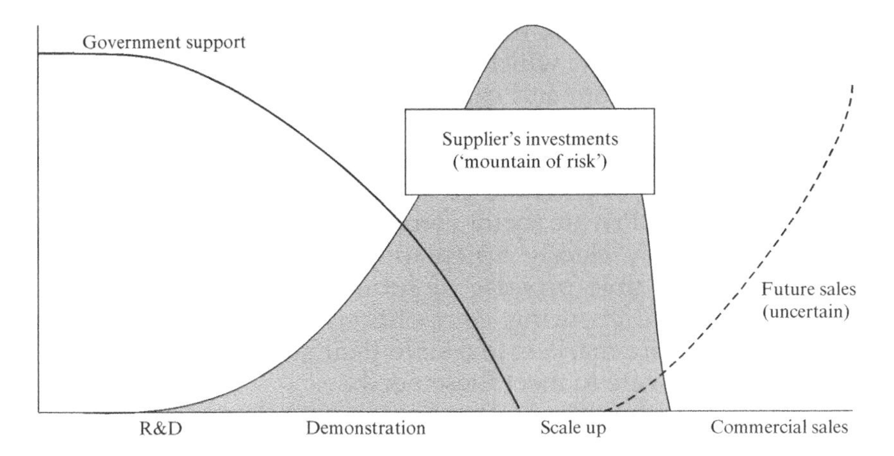

Note: Shown in the figure is the risk profile of a supplier of innovation. The time of maximum risk (indicated as the 'mountain of risk') occurs during demonstration and scaling up. Suppliers have to invest while there is uncertainty about future commercial sales.

Source: Adapted from DTI (2006).

Figure 5.1 EIAG reimagined the 'valley of death' as a 'mountain of risk'

by Lord Sainsbury, then UK Minister for Science and Innovation, in 2003 (DTI, 2006) under the chairmanship of Dr J.C. Frost (OBE). EIAG was set up to examine the relative failure of the environmental sector to bring innovations to market and to make recommendations on how this could be remedied.

EIAG concluded that rather than a funding gap producing a 'valley of death' between the relative low-cost and low-risk R&D phase of product development and the revenue flow in the commercialization phase, suppliers faced a 'mountain of risk' (see Figure 5.1). They identified the uncertainty of future sales, rather than the lack of funds, as the dominant barrier and determined that one of the root problems of the relative failure of environmental innovation was a lack of 'credible articulated demand' rather than any lack of research, invention or innovative aspiration. EIAG felt that this could be remedied by the government taking action to mobilize the supply chain to deliver environmental innovations through intelligent supply chain management.

It was on this basis that EIAG conceived, developed and tested a methodology, which could be capable of driving environmental innovation, and called it 'forward commitment procurement' (FCP). FCP sought to apply, in the public sector, the approach taken by business in using supply chain management and procurement to promote investment in new products along its supply chain. Private sector companies actively manage their supply chains by clearly articulating their future needs and providing a credible promise of future sale (a 'forward commitment'), while ensuring competition, to provide sufficient certainty in future markets to enable their suppliers to compete, invest and innovate to meet these needs.

EIAG described the FCP approach as follows: 'In essence, the approach involves providing advance information of future needs, searching out and engaging with potential suppliers and, critically, incentivising them through a Forward Commitment – the promise of current and future business to promote investment in innovative new product development' (DTI, 2006). Later publications state explicitly that the FCP method was designed to manage the risks involved in innovation procurement (Department for Business, Innovation and Skills, 2011).

EIAG undertook to develop and demonstrate the FCP methodology. This first demonstration project concerned a problem that HM Prison Service had with the disposal to landfill of more than 40 000 foam mattresses annually. Adopting the FCP method led to a fundamental shift in the procurement approach. The resulting Zero Waste Mattress service eliminated waste to landfill and projected savings approaching £5 million over the life of the four-year contract (Department for Business, Innovation and Skills, 2009).

The FCP methodology was further developed by the UK Department for Business, Innovation and Skills (BIS), and was adopted in 2009 as the methodology for delivery of innovation procurement in pilot projects within the LCB-Healthcare Project. This enabled the FCP method to be further tested through four pilot projects in the UK, the Netherlands and Poland. It also offered the opportunity to examine the transferability of FCP to other European Union (EU) member states. Three of these pilot projects form the basis of the research discussed in this chapter.

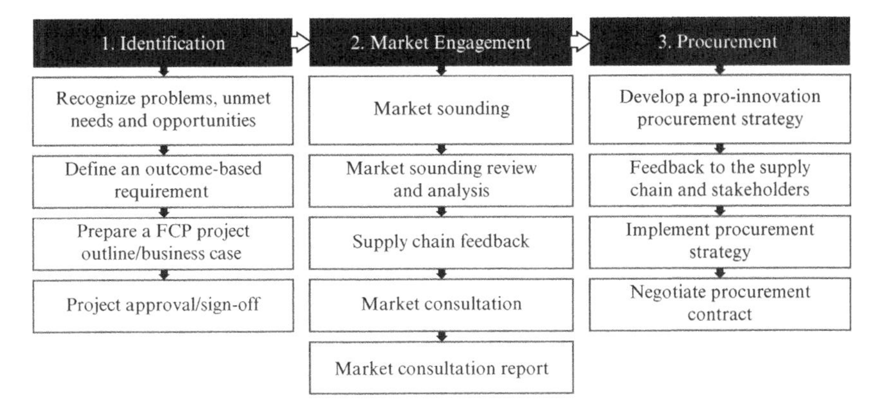

Figure 5.2 Steps of the forward commitment procurement method according to the Department for Business, Innovation and Skills (2011)

The FCP Methodology

The Department for Business, Innovation and Skills (2011) has published a guide that describes the various steps in the FCP method. In this guide, FCP is presented as a process consisting of three phases, each consisting of a number of specific steps (Figure 5.2).

FCP provides a framework of actions that together can change the market situation to one that supports the delivery of the outcomes the customer requires. The three phases of the FCP method are further discussed in the next paragraphs. Special attention is paid to the second phase, the market engagement phase.

First phase of the FCP method: identification

The FCP method starts with an identification phase in which the customer identifies unmet needs within the organization that requires innovation. Alternatively, the customer can also identify opportunities (such as an investment in refurbishment or re-tendering of a major contract) where innovation could add value. The requirement is then defined and formulated in terms of the outcomes needed. In any case, FCP is applied by the end-user of the solution that is provided. Therefore, FCP can be characterized as a form of direct PPI (see Edquist and Zabala-Iturriagagoitia, 2012). The first phase of FCP coincides with the two first stages of a PPI process as described by Edquist and Zabala-Iturriagagoitia (2012); see also Chapter 1 in this volume.

An FCP project outline and business case is prepared and, after approval, the project moves on to a market engagement phase. In this phase, the customer can interact and exchange information with potential suppliers.

Second phase of the FCP method: market engagement

Usually, the market engagement phase has two stages: a market sounding stage, and a consultation stage. This market engagement phase is 'pre-procurement', that is, it takes place before a formal procurement procedure begins.

A key characteristic of the FCP method is the market engagement phase in which clients take action to create the market conditions necessary to support the delivery of innovative solutions by the supply chain (Department for Business, Innovation and Skills, 2011). Although the research focuses particularly on the market engagement phase and its role in managing risk, this does not mean that other phases are not considered important in this regard. The identification phase is critically important in managing risk through such actions as end-user engagement, securing leadership support and, most fundamentally, the identification of a genuine unmet need and an accurate definition of the requirement (LCB-Healthcare, 2012).

Third phase of the FCP method: procurement

The development of a procurement strategy and its execution comprise the third phase of the FCP method. It coincides with the third and fourth stages[1] of the PPI process of Edquist and Zabala-Iturriagagoitia (2012). The procurement strategy is developed after the market engagement phase has concluded, in order that the information gained through the market engagement can be used to develop the most suitable procurement strategy.

Actions Taken within the FCP Stages

In analysing the different phases of FCP, the chapter identifies some distinct actions undertaken within the different steps of FCP, each of which requires one or multiple actions, that is, goal-oriented purposeful human actions to reach goals set. The chapter then goes on to examine (the implementation of) FCP in the case studies on the basis of these actions. Hence, the most elemental unit of analysis in the research is represented by actions of FCP. The purpose of each action is determined first (see below). During the case studies the impact of these actions on risk is assessed, using the theoretical framework of agency theory and transaction cost economics.

Actions identified within the market engagement phase of FCP

The following actions are identified as part of the market engagement phase, based on a desk review of documentation on FCP and available case study documentation. It is important to note that these actions can be taken a number of times throughout the FCP market engagement process and that FCP activities may comprise several actions.

Communication Communication is the act of conveying information. In this chapter, communication is seen as the transferring of information from the customer to the supply chain.[2] This is also called disclosure (Winch, 2002). Von Hippel (1986) suggests that an accurate understanding of user needs is essential for the development of commercially successful products. Edler and Georghiou (2007) furthermore state that it is important that suppliers are given early signals regarding future public demands.

Sounding Market sounding provides a framework and opportunity for potential suppliers to respond to the requirements of the customer. Winch (2002) calls this form of information transfer feedback. Although the customer has identified and formulated its requirements, it does not know if these requirements can be satisfied by the market. Sounding enables the customer to check whether suppliers are able to deliver the required outcomes, whether their requirements are formulated appropriately, and whether suppliers are interested in the upcoming procurement project. The feedback from the market sounding phase enables the customer to assess the capacity, capability and willingness of the supply chain to deliver a solution based on the information it has provided. It also enables the supply chain to comment on the requirement which will contribute to the customers' refinement of the requirement. Aligning customers' needs and market capabilities is an important factor in the success of PPI (Edler and Georghiou, 2007). If alignment is necessary (public demand and market capabilities do not match), the project, approach or requirement may need to be refined (see 'Refining the requirement', below).

Sounding may also provide public procurers with information about other required actions. For instance, information may be gained about regulatory and other barriers that suppliers face, and help to determine the procurement strategy that is most likely to deliver the desired outcomes.

Refining the requirement, or market demand In the first stage of market engagement the market is informed about an upcoming project (communication), and is asked to respond to a number of questions (sounding). The sounding informs customers of the market situation and, on the basis of this information, customers may adjust the scope of the project or adjust their requirements. This helps to reduce any gap between customer needs and market capabilities. If this gap is too large, innovation may not be feasible (Edler et al., 2006) or will take too long.

Signposting demand Signposting a larger future market is a way of increasing market pull by indicating the availability of a wider market, and helping suppliers to assess whether the market is of sufficient scale to warrant their investment. This is not to be confused with joint procurement, where demand is aggregated in the form of joint purchase. There are disadvantages associated with joint procurement. Examples are an increased complexity of the purchasing process and loss of flexibility and control (Schotanus, 2007), and potentially a 'dumbing down' of the requirement to the lowest common denominator. In determining the applicability and value (for example, economies of scale) of such formal aggregation of demand, the needs, requirements or specifications need to be similar among procuring agencies (Schotanus, 2005). This is much easier when procuring a commodity than in cases where innovation is required. In many cases, the value of aggregating demand is overstated. The difficulty and complexity of a formal aggregation may outweigh the benefit and arguably increase risk. In many cases simply signposting larger demand is adequate as sophisticated suppliers can anticipate the advantage of being first mover into a nascent market.[3]

Facilitating networking In PPI, it may be necessary for several supply-side organizations to cooperate and jointly develop a new product or service to meet the customer's requirements. By facilitating networking the public sector organization attempts to stimulate networking among different suppliers and across supply chains and supply sectors. Edquist and Hommen (1999) state that firms almost never innovate in isolation. Interorganizational cooperation thus enables innovation. Such cooperation can be forced (Bossink, 2007), as physical and human resources, subsystems, components, technologies, skills, information and knowledge can be dispersed among various organizations (Rutten et al., 2009). The first steps in cooperation concern discovering and exploring opportunities (Bossink, 2007). FCP aims to create the opportunity (the requirement) and then encourage cooperation through facilitating networking amongst and between supply chains.

Aligning the procurement strategy A pro-innovation procurement strategy sets out a procurement pathway designed to support and enable innovation and achievement of the required outcomes through the procurement process. Several choices have to be made, for example regarding the procurement procedure (such as competitive dialogue), selection criteria and award criteria. The strategy can also be used to set out attitudinal parameters to align the approach taken by the team engaged in the procurement process. The procurement strategy is developed after the market engagement phase is concluded, in order that the information gained through the market engagement can be used to develop the most suitable procurement strategy.

THEORETICAL FRAMEWORK USED TO EXAMINE THE IMPACT OF FCP ON RISK

Having distinguished the different actions undertaken in FCP, agency theory and transaction cost economics (TCE) are used as the theoretical framework to structure the analysis of how these FCP actions influence risk in each of the case studies, and to gain insights into how public sector customers manage perceived risks in an FCP project. They are selected because both agency theory and TCE discuss how risk influences the behaviour of actors (in this case, customers and their potential suppliers) and discuss this behaviour in the context of transactions (in other words, procurement).

Agency Theory and Transaction Cost Economics

Agency theory concerns the relationship in which one party (the principal) delegates work to another (the agent), who performs that work (Eisenhardt, 1989). Agency theory is concerned with resolving two problems: (1) that principals and agents have different (conflicting) goals; and (2) that principals and agents have different attitudes towards risk-taking. Agency theory is similar to TCE (Eisenhardt, 1989), although the former emphasizes that actors have different goals and risk attitudes.

TCE is also referred to as the economics of contracting. A contract will concern a relationship between principal and agent (Eisenhardt, 1989). TCE supports two behavioural propositions: bounded rationality and opportunism (Williamson, 1981). These two behaviours are also assumed in agency theory (Eisenhardt, 1989). Bounded rationality implies rational decision-making by buyers and sellers under conditions of incomplete information (Parker and Hartley, 2003). Information asymmetry is when principal and agent do not possess the same information (Tysseland, 2008), which can lead to problems arising from opportunistic behaviours of the parties, that is, opportunism. Opportunism refers to the seeking of self-interest with guile (Williamson, 1979).

According to Williamson (1979, 1981), the three critical dimensions for characterizing transactions are: (1) uncertainty; (2) transaction bounded investments; and (3) the frequency in which transactions recur. In relation to the context of a single procurement project (for example, single transaction), which is what we are considering in this research, the aspects of uncertainty and transaction bounded investments are most relevant. This leads us to consider risk in terms of uncertainty and transaction bounded investments.

Uncertainty

Information determines an actor's ability to accurately estimate outcomes. The absence of information leads to uncertainty (Edler et al., 2006). A natural response to uncertainty is information-seeking behaviour. However, seeking out information can be difficult and expensive (Eisenhardt, 1989).

Transaction bounded investments

Transaction bounded investments are inputs (for example, time, money) that are specifically used for a single transaction and are therefore not usable for other uses without loss of value (Dorée, 1996). Cowen and Parker state that transaction costs 'arise from seeking out buyers and sellers and arranging, policing and enforcing agreements or contracts in a world of imperfect information' (Cowen and Parker, 1997, p. 37, cited in Parker and Hartley, 2003). Such activities are often costly (Coase, 1960).

As procuring innovations requires more complex procurement processes, it is likely that the transaction bounded investments will be higher compared to cases where additional procurement-related activities and development are not required (Hommen and Rolfstam, 2009). As risk management processes consume valuable resources and can themselves constitute a risk to the project that must be managed effectively (Chapman and Ward, 2003), we can therefore expect that any risk management activity (such as FCP) will incur transaction bounded investments, that is, from undertaking the actions defined earlier. There are two approaches to influencing transaction bounded investments. One course of action entails making the inputs of the transaction less bounded to a single transaction (the action of signposting demand is an example). Another approach concerns influencing the effort required for the actions that incur transaction costs. For instance, clients can reimburse private organizations for certain development costs, hence transferring transaction bounded investments from market party to customer.

Applying the Theoretical Framework to the FCP Method

The level of uncertainty and the level of transaction bounded investments influence the course of transactions – that is, with increasing levels of uncertainty transactions become less attractive – and any investment in the transaction has to demonstrate a return proportionate to the risk and reward. Therefore, reducing uncertainty and being able to determine the value of investing time and money in a transaction will contribute to the management of that risk.

This chapter therefore assesses the effect of the FCP method on perceived risks by looking at how FCP influences the levels of uncertainty and transaction bounded investments in each of the case examples. As the FCP method aims to help to manage risks for both customer and supplier, applying FCP should result in decreased levels of uncertainty, and the ability of both parties to assess whether the levels of transaction bounded investments will reap a reward proportionate to the level of investment.

The FCP methodology provides an incremental framework, meaning that both customers and suppliers can approach innovation procurement in a staged process, whereby at each stage the level of uncertainty and transaction bounded investments can be assessed as manageable or unmanageable and a decision made to proceed or withdraw. It is reassuring for both parties to know that they can withdraw at any stage in the process, having incurred only the justifiable opportunity costs.

DESCRIPTION OF THE CASE STUDIES

The three case studies chosen for the study are pilot projects undertaken as part of the LCB-Healthcare project and involve two hospitals from the UK National Health Service (NHS) and an academic hospital in the Netherlands. The approach in data collection is based on Makkonen et al. (2012) who discuss a narrative approach. Narratives provide rich information on time, place, actors and context that may be crucial in interpreting events. Several steps were undertaken in the data collection phase, going from studying relevant documentation related to each project (external and internal), to having interviews with multiple members of each project's team. These sources are supplemented with direct or participant observation when possible. Based on this information resulting from multiple sources, a final narrative has been constructed for each case study (summarized in Table 5.1).

Bed Washing Innovation: Erasmus University Medical Centre (Erasmus MC), the Netherlands

Erasmus University Medical Centre (Erasmus MC) needed to renew its bed washing facility to provide enough clean and disinfected beds for its daily operational needs. This currently exceeds 70 000 beds per annum and is expected to increase. The current machine is nearing the end of its life, is labour

Table 5.1 Overview of case studies

Organization	Project	Progress to outcome	Status
Erasmus Medical Centre (the Netherlands)	Cleaning of 70 000 hospital beds per annum	New product developed	Contract signed subject to successful demonstration
Rotherham NHS Foundation Trust (UK)	Ultra-efficient lighting for future wards	New product developed	Memorandum of Understanding (MOU) signed with preferred supplier
Nottingham University Hospitals NHS Trust (UK)	Integrated ultra low carbon energy solution for the city campus site	New to market products and services forthcoming	Refined specification drafted based on market engagement Feasibility study of innovative technology in progress

intensive and uses a large volume of water and energy to operate. Beginning the FCP project more than three years before the solution was needed gave the hospital time and opportunity to explore the possibility of procuring a more efficient, more effective and sustainable solution through PPI.

Identification

A cross-departmental project team wanted to allow time for suppliers to come up with new ideas and build consortia so they began the project much earlier than normal. The joint procurement and operations project team defined both the characteristics of the current process (including the current carbon footprint of the process) and the required outcomes that they were looking for.

Market engagement

A market engagement plan was developed and the project team began the market sounding exercise in September 2011. A short survey confirmed supply chain interest in the market consultation and the procurement opportunity. The project team also explored the wider market demand for the solution and they identified at least six other hospitals in the Netherlands that were interested in the outcomes of the consultation. The market consultation stage was concluded at the end of 2011 and was followed by a 'market meeting day' in January 2012.

The goal of the market meeting day was to discuss the challenge in detail: how to efficiently clean beds in a sustainable way in the context of Erasmus MC. A large number of market parties from a wide range of backgrounds (around 60 people) attended, as well as stakeholders and several other healthcare organizations.

The market meeting day gave the Erasmus MC project team confidence that there were both interest and capabilities in the supply chain to deliver the specified outcomes.

Procurement

On the basis of the information from the market engagement phase, Erasmus MC developed a pro-innovation procurement strategy and established three award criteria for the new bed cleaning facility:

1. total cost of ownership/service;
2. carbon footprint; and
3. fit with strategy of Erasmus MC organization.

Making carbon footprint an award criteria sent a strong signal to the market that suppliers have an important role to play in reducing operational and embedded carbon in their products and their supply chain.

Following evaluation of the pre-qualification questionnaire, eight parties or consortia were invited to present their initial ideas and thoughts during the first round of the competitive dialogue. Through further dialogue rounds, in which ideas were transformed into concrete solutions, two parties were invited to submit final offers in December 2012.

In June 2013, one party was contracted to provide the solution that is innovative in the way it cleans hospital beds and mattresses, economically and environmentally sustainable, and can be managed by the hospital. The solution was provided by a Dutch small enterprise that combined robot technology from the car manufacturing industry with a patented steam nozzle.

Ultra Efficient Lighting for Future Wards: The Rotherham NHS Foundation Trust, UK

The FCP project undertaken by the Trust set out to deliver a step change in the patient experience and a step change in the efficiency and effectiveness of hospital ward lighting, in a way that was cost-effective, 'future ready' and transferable to other NHS hospitals.

Identification

The Trust needs to deliver carbon reductions, save money and at the same time deliver excellent services to patients. The opportunity for innovation was presented by an eight-year refurbishment programme. Working with the BIS/DH (Department of Health) FCP programme manager, and with the vision of the chief executive officer (CEO) for a 'Hospital of the Future' firmly in their minds and adopting FCP thinking, the project team set out to define what they needed in terms of outcomes, rather than in terms of the products that were available on the market.

Market engagement

Once identified, this 'unmet need' was communicated to the supply chain in outcome terms as part of a market sounding exercise which was launched via a prior information notice in the *Official Journal of the European Union* (*OJEU*) more than two years before the solution would be needed on site. The main vehicle for communication with the supply chain was a market sounding prospectus which set out the context, drivers and requirements and scale of the procurement. The market engagement process was carried out in two stages: a market sounding where interested parties completed a short market sounding response form, and a market consultation workshop. The response from the supply chain was enthusiastic and demonstrated that there was motivation, capacity and capability in the supply chain to deliver a solution. The level of interest and information that the Trust received from the supply chain during the market engagement phase enabled it to refine the outcome-based specification, develop a pro-innovation procurement strategy and to proceed to the procurement phase with confidence.

By providing advance information on the requirement, in the context of a major procurement, using an open, outcome-based specification and by stimulating cross-supply chain cooperation, the Trust aimed to give the supply chain the time, opportunity and motivation to come up with an innovative solution.

Procurement

The project team developed a pro-innovation procurement strategy that was approved by the Trust Board. The strategy included for example a revised outcome-based specification based on the market response, use of a revised pre-qualification questionnaire competitive dialogue, and whole-life costing. The strategy also served to align the project team as they entered dialogue and negotiation.

Stimulated by the advance warning of the forthcoming procurement and enabled through the consultation process, one pan-European consortium worked together in advance of the tender and was subsequently well prepared to come forward with an innovative solution that met, and indeed exceeded, the Trust's expectations. The pro-innovation procurement approach has brought to the market an integrated 'future ward' modular solution, with integrated biodynamic lighting, trunking, and storage for use by patients and staff. Detailed costings, verified by an independent quantity surveyor, show that the innovative solution will cost the same as a standard ward solution, with not only the required step-change in patient experience and lighting efficiency, but also with reduced on-site build time and additional benefits.

An Integrated Ultra Low Carbon Energy Solution: Nottingham University Hospitals NHS Trust, UK

Nottingham University Hospitals NHS Trust (NUH) is one of the largest acute trusts in the UK. For the last 35 years the primary source of heat for one of its locations (City Site) has been a coal-fired boiler which is now coming to the end of its useful life. The director of estates and facilities saw this as an opportunity to fundamentally rethink the Trust's approach to energy and reduce the hospital's carbon intensity, and in doing so reduce exposure to rising energy prices, the pricing of carbon and the impact of future low carbon legislation.

Identification

Initially the Trust commissioned consultants to advise on the replacement of the site's energy systems. The resulting recommendations, based primarily around large centralized energy generation on site, would deliver a significant one-time reduction in energy use and emissions, but would not necessarily achieve the long-term flexibility, carbon reductions and reduced exposure to rising energy prices that the Trust required. The project team agreed that new, more efficient combined heat and power systems would be only part of the solution, and in isolation would not deliver the outcomes they were looking for. The outcome the Trust wanted was to achieve more than a step-change reduction in energy demand and emissions, and to reverse historical trends and deliver continuous efficiency improvements across the Trust over the next 20 years. An innovative response was needed from the supply chain, and this meant that the Trust needed a new procurement approach.

Market engagement

The market sounding was advertised in the *OJEU* in November 2011, and the call for innovative solutions was proactively communicated to the market. It set out to get feedback from potential suppliers through a market sounding exercise. A number of questions were put to potential suppliers, and the market sounding process received an enthusiastic response from the supply chain, with more than 120 people attending a site visit in January 2012. More than 65 high-quality responses to a market sounding questionnaire were received. These responses gave a highly informative snapshot of the state of the market.

Procurement strategy

The feedback from the supply chain has been invaluable in helping the project team to outline the options available to the Trust, and these are now under consideration. This includes new-to-market solutions ripe for early adoption. The market sounding and supplier engagement has led to the Trust rethinking its approach to energy provision and energy efficiency. In the feedback to the market sounding, the supply chain expressed a need for the Trust to determine in more detail the nature of the energy solution for the site before beginning a formal tendering process. The Trust therefore undertook wider consultation and internal discussions, and concluded that a distributed energy solution that used the most efficient technology available would be the best way to deliver the required outcomes. On this basis a feasibility study was undertaken as part of a business case and procurement strategy. At the time of writing in December 2013, the feasibility study was in progress.

RESULTS FROM THE CASE STUDIES

In this section, the results from the case studies are jointly discussed. First, the drivers for innovation and perceived risks and uncertainties are presented. Next, the actions taken during each of the three phases of the FCP method are described. The effects of these actions are then discussed.

Drivers for Innovation

The drivers behind the need to innovate and the adoption of new procurement approaches that supported innovation were similar across the three case studies: a combination of operational needs, environmental sustainability targets, carbon reduction and cost saving. In common with all public healthcare agencies the case study subjects faced national or service-wide carbon reduction targets, rising costs of energy and, in some cases, financial penalties relating to carbon emissions. It can be argued that these drivers are not just common to public healthcare agencies, but common to the entire public sector.

Although there were several drivers for innovation, it is unlikely that the projects would have begun without the intervention of partner organizations. For example the Department of Health provided funding to carry out baseline energy studies and, through the LCB-Healthcare project, the Department of Health, the Department for Business, Innovation Skills and the Netherlands Organization for Applied Scientific Research (Nederlandse Organisatie voor Toegepast Natuurwetenschappelijk Onderzoek, or TNO) provided project support, facilitation and know-how. Prior to engagement with partner organizations the healthcare organizations had no awareness or experience of innovation procurement as a mechanism to deliver better outcomes.

Perceived Risks and Uncertainties

The customer's perception of risk fell into two categories. The first category related to uncertainty about the delivery and performance of the solution. Will the market deliver and will the solution work? The customer was uncertain about available technologies, market capabilities, market interest in and awareness of the project, the ability of the solution to meet the specified outcomes, reliability of the solution, timeframe in which the solution could be provided and the suitability of the procurement strategy.

The second category related to uncertainty about the viability of the outcome. Can we use the solution? Will it fit and be accepted into the organization of the hospital? Will it be 'buyable' as opposed to 'worth it'?

The uncertainty from the supplier's perspective came down to 'Will they buy it (even if it meets the requirements)?' A supplier's uncertainty needs to be reduced by better understanding of the customer's needs and confidence that the customer is a credible buyer and that there is a wider future market here. Reducing the level of uncertainty in these areas therefore leads to a reduced perception of risk, and the ability to assess whether it is worthwhile to invest in the transaction and, if so, what the level of that investment should be.

Actions Undertaken in the FCP Pilot Projects

The LCB-Healthcare pilot projects followed the FCP method as proposed by the Department for Business, Innovation and Skills guide (see Figure 5.2). This guide provides insights into and examples of how to implement each of these steps and activities. As mentioned previously, the research focused on the market engagement phase of the FCP process.

First stage of market engagement phase: communication and sounding

Following the completion of the identification phase, in all the case examples studied, the project teams embarked on a market sounding exercise to communicate the requirement to the market and gather information from the supply chain.

Purpose In communicating the requirement to the market, the customer aimed to reduce the uncertainty of the supplier about the requirements of the customer and wider market. In gathering information from the supply chain, the customer aimed to reduce uncertainty about the ability of the supply chain to deliver a solution.

Steps The first action was the development of a market sounding prospectus (MSP). The MSP set out the context, drivers, requirement and procurement timeframe and invited the supply chain to respond in writing.

The market soundings were launched via a prior information notice in the *OJEU*. Other communication actions included direct emails, telephone calls, articles and adverts in trade journals, use of industry and innovation networks, personal networks, and so on, to ensure that suppliers were aware of the project.

In two cases (Rotherham and Erasmus MC), the customer communicated with other hospitals to inquire about their interest in these projects. In both cases, there was sufficient interest from other hospitals to signpost this future market in the MSP.

In one case (Nottingham), due to the nature of the project, the market sounding process involved direct contact with suppliers in the form of a site visit day. This enabled interested suppliers to visit the site in advance of responding in writing to the market sounding. This event also provided an opportunity for suppliers to gather more information about the requirement, assess the credibility of the customer and network with other suppliers.

In each of the pilot projects, the market sounding was facilitated by the use of response forms in which suppliers were requested to respond to the MSP and answer some questions.

Perceived actions Communicating, sounding, signposting demand, facilitating networking (one case).

Second stage of market engagement phase: consultation

This stage involved personal contact with suppliers that had responded to the market sounding.

Purpose In deepening the two-way contact between supplier and customer, the aim is to continue transfer of information to reduce uncertainty and enable both parties to assess the value in continuing with the process.

Steps A market consultation workshop (Rotherham) or market meeting day (Erasmus MC) created a two-way platform in which customers and suppliers exchanged information and enabled suppliers to network and make connections. This process of networking was further enabled by the publication of a 'Company Directory' containing contact information on participating suppliers. These events were similar to the site visit day at Nottingham University Hospitals, although the Nottingham event was organized prior to the deadline for responding to the market sounding.

Perceived actions Communicating, sounding, facilitating networking.

Impact of market engagement

In all projects an analysis of the market sounding information was carried out and reports prepared for senior management. In all cases the analysis served to reassure the customer and reduce uncertainty. The level of interest shown by suppliers, and the information they provided, enabled the healthcare agencies to better assess the market situation, reduced their uncertainty and gave them confidence to continue to the second stage of market engagement.

The project team at Nottingham carried out an analysis of the responses from the market, which enabled them to prepare a progress report to a finance management committee and get approval to proceed to the next stage of the project on the basis of a refined project scope and approach. The overall results of these market engagement activities were that risks were managed to such an extent that customers were confident that the project would deliver a desirable outcome.

Using the results of the market engagement phase: refining outcomes, aligning procurement strategy

During the period of the study, two of the case studies had progressed to the third stage of the FCP process and had developed a pro-innovation procurement strategy.

Purpose The purpose of alignment was to understand and use the information exchanged in the market engagement phase to optimize the chances of success of the procurement in delivering the required outcomes. The supplier also underwent a process of alignment of their goods and services or investments to match the customer's requirement where it was convinced that it was worth the investment.

Steps The information gleaned from the market engagement activities directed the development of the procurement strategies. Basing the procurement strategy on knowledge about the supply chain further increased the confidence of the customer that the outcomes they sought were deliverable, and that the strategy was aligned to the market situation. Making a public version of the procurement strategy available (Rotherham) helped to reassure suppliers that the customer was listening to them and was serious about buying a solution.

In the case of Rotherham, the alignment process included refinement of the outcome-based specification and changing the basis for measuring lighting based on the information and advice received from suppliers during the market engagement phase. It also led them to change their standard pre-qualification questionnaire in several ways, for example to ensure that suppliers with no or limited experience of delivering to the healthcare market were not excluded from the competitive dialogue phase. The focus at Erasmus MC was to make sure that the procurement procedure did not hinder participation in the competitive dialogue, ensuring that potential solutions would not be excluded unnecessarily.

During the market engagement phase, it became clear that many suppliers were unfamiliar with public procurement procedures and also with award criteria like carbon footprint or total cost of ownership. The procurement strategy was aligned to this information by adopting award criteria that included only a minimum set of requirements in order to allow equal access to competitive dialogue.

Another alignment resulted from suppliers' concerns over confidentiality and investment: Erasmus MC stated that suppliers would not have to enter the dialogue phase with detailed proposals, but only with initial ideas. Development of the solutions beyond initial ideas that would incur greater development costs could then be postponed until subsequent dialogue rounds at which point some of the parties would have been excluded from the tender process. This is an example of how a customer has helped to manage a supplier's risk by making transaction investments incremental.

In another alignment action the team responded to the lack of familiarity with total cost of ownership and carbon footprint calculations by providing technical assistance to suppliers via its partnership with TNO.

Perceived actions Refining requirements, aligning procurement strategy.

Assessment of Individual FCP Actions

In this section, the effects of FCP are assessed by looking at the impact of the actions taken on the level of uncertainty and transaction bounded investments.

Communication

By providing suppliers with accurate and well-considered information (based in the identification stage) not only on their requirements but also on the context of and drivers for the project, customers reduced the level of uncertainty for suppliers regarding the customer's needs, necessity for innovation, commitment to procure and capacity for innovation. However, providing this information required the customers to make transaction bounded investments, for example in terms of time resources.

Sounding

Sounding was done on multiple occasions and by different means. The effects were similar, however. As customers acquired information, their level of uncertainty decreased. Customers reported that they were more certain that suppliers could and would deliver desirable solutions. Hence suppliers reduced the customer's level of uncertainty by responding to the customer's requirement and communicating their motivation, capacity and capability. As suppliers engaged further with customers, their level of uncertainty decreased. However, this action incurred costs for both customers and suppliers, since providing this information required the customers to make transaction bounded investments, for example in terms of time resources. Suppliers incurred costs when they responded to the customer's sounding. However, sounding enabled both parties to assess the value of continuing to engage with the FCP process.

Refining outcomes (demand)

In the case of Rotherham, the outcome-based requirement that was communicated to the market was refined and simplified on the basis of the market engagement process, and this refined outcome specification formed the basis for the tender. In the case of Nottingham, the project scope was initially deliberately broad to enable the Trust to investigate all possible approaches. After reviewing market responses, the requirement remained the same, but the Trust adjusted the demand by limiting the next phase of market engagement to considering only a decentralized, rather than centralized, energy solution.

Signposting wider demand

This action occurred in two cases. In these cases, potential routes to a wider market were explicitly mentioned. By demonstrating a wider market, both the Rotherham and Rotterdam cases indicated that the likelihood of future sales was greater than the initial customer. In the Rotherham case it was an explicit requirement that the solutions should be transferable across the NHS. This signposting of wider demand supported suppliers in their market assessment and opportunity assessment, enabling them to assess the value of continued transaction investment. For example, it showed that their investment was not bound to a single transaction. Knowing that other healthcare agencies shared their unmet needs and requirements increased the customers' confidence and reduced the level of uncertainty. It demonstrated that their identification process had credibility and that they could benefit from the existence of a wider market demand and its impact, that is, increased market pull.

The informal approach to aggregating demand, which involved consulting colleagues in the healthcare sector regarding their potential as future buyers, required marginal effort on the part of the customer, yet had the desired effect of demonstrating the wider market demand, thereby reducing uncertainty on the part of suppliers. It also avoided the risks that are incurred in joint procurement actions. Where customers seek to formally aggregate demand in joint procurement, the level of investment is considerably higher and additional risks arise.

It can also be argued that innovation procurement projects in the healthcare sector have the potential to benefit from an implicit scaling-up of demand that results from a high degree of commonality of healthcare requirements across Europe and indeed globally; that is, the unmet needs of one hospital in the UK are likely to be similar to the requirements of the other 15 000 hospitals in Europe. By expressing an unmet need and signposting wider interest, the supply chain can extrapolate sales to a much wider market.

Facilitating networking

All the case studies facilitated supply chain interaction. The effects of this action are not always visible to the customer. There are, however, some examples from the UK case studies. One consortium that responded to the Rotherham tender consolidated at the market consultation workshop. This pan-European consortium subsequently developed a new product, and the two lead partners formed a small firm to commercialize the solution; connections made at the NUH site visit led to cooperation between a small provider of an innovative technological solution (fuel cell combined heat and power) and a P21+ framework contractor.[4]

Aligning procurement procedure

Only two case studies have entered the procurement stage of the FCP process. In these cases, the alignment decreased the clients' level of uncertainty as the procurement strategy was tailor-made for delivering a desirable solution. In both cases the formalization and organizational approval of a pro-innovation procurement strategy reduced the uncertainty as the organization entered the formal procurement phase.

Overall Effect of Applying the FCP Method

The research shows that the FCP approach led to reduced uncertainty on the part of the customer and supplier, enabling them to progress from information exchange to procurement of innovation. In two cases the process led to the development of new products, and in one case the (anticipated) early adoption of new to market products and technologies. FCP therefore seems to be capable of managing customers' perceived risk and overcoming the risk-adverse reputation of public sector buyers. Both suppliers and customers agree that these developments would not have been possible without FCP.

Some suppliers demonstrated uncertainty arising from lack of familiarity with public sector customers asking for innovative solutions in the market engagement process and competitive dialogue processes. These were largely managed through the alignment of the procurement strategy.

The research indicates that the level of transaction bound investments (resources) needed to implement the FCP process on the part of the customer is higher than for a 'standard' procurement, that is, more actions are needed by a larger group of people, and in all cases the project was supported by external agents. The level of pre-procurement transaction-bound investment on the part of the supplier was also higher than a standard procurement.

However, the notable fact is that both customers and (interested) suppliers continued to engage in the FCP process. This suggests that the value of the FCP process in reducing uncertainty and perceived risk was worth their investment. At any point where this was not the case, the customer could withdraw from the process and suppliers could decide that this was not for them or proceed with minimal investment (with an existing and proven product, for example).

CONCLUSIONS AND FINAL REMARKS

Based on this study, forward commitment procurement has been shown to provide a practical and effective methodology to undertake PPI projects. Applying the FCP method reduced the risks perceived in the PPI case studies discussed. In all cases the actions taken positively affected risks, although taking action required additional (but justifiable) investment. The process demonstrated how customers and suppliers can manage each other´s risks through a process of engagement.

The FCP method helps to manage perceived risks in innovation procurement projects, but its application, like any risk managing process and any innovation process, requires investments on the part of the customer and suppliers. The market engagement actions play an important part in the effectiveness of the FCP method, but the importance of the identification phase in determining and bringing unmet needs into focus, and articulating accurately the customer requirement, underpins successful market engagement.

Innovation takes time. Market engagement therefore needs to begin well in advance of the time that the required good or service is required. In all these case studies, action began years ahead of the solution being required. This allows enough time for customers, but more especially suppliers, to take the required steps towards the successful delivery of innovation.

The case studies provide valuable insights into how customers can reorient their procurement approach to become more ambitious and innovative; how customers and suppliers can communicate and engage effectively to bring new innovations into the market; and how customers can buy something innovative in a way that does not appear too risky.

The range of case examples examined here (and other experiences with applying the FCP method) leads us to conclude that the FCP method is widely applicable: although originally developed in the UK, it has now been applied in other member states, and to a variety of projects.

With ever increasing societal challenges, public organizations will increasingly need to innovate or risk failure. This will require organizational change, both in the way that innovation is perceived and supported, and in the way that procurement functions are managed. To support innovation, public procurement needs to become more strategic and less commodity based.

A lack of awareness, capacity and capability as well as perceived risks is currently preventing organizations from undertaking PPI projects. Although the adoption of FCP is limited, these case examples indicate that it provides a practical and effective methodology whereby public sector organizations can buy innovative solutions, with associated benefits for both the organization and society.

Limitations and Suggestions for Further Research

The study and pilot projects undertaken have shown that the FCP process enables innovation procurement in healthcare organizations. The authors acknowledge the limited scope of this research. The research focused specifically on the effect of the market engagement phase on the risks and perceived risks of innovation procurement on the part of customers and suppliers. We have to acknowledge that all stages of the FCP process have an impact on risk which, although apparent in the study, were not analysed.

The number of FCP projects remains limited and more

case studies would be valuable in increasing the current body of knowledge. Further research would help to refine the application of FCP, and assist public sector organizations to deliver the outcomes that they and society need to meet the social, economic and environmental challenges of the future.

PPI projects need to stem from genuine unmet needs, yet few organizations have the mechanisms in place to define their future needs, and fewer see procurement as a strategic tool to meet these needs. A study of the identification phase would inform this process. Linked to this, examining different ways of maximizing the return on the transaction investments incurred by customers and suppliers in implementing an effective PPI project would be an interesting and valuable addition to the research in this field.

ACKNOWLEDGEMENTS

The authors would like to express their gratitude to the partners and pilot project participants of the LCB-Healthcare Network. The LCB-Healthcare website contains further information about these and other case studies. Special thanks go to Dr J.C. Frost for his comments and suggestions.

NOTES

1. Tendering process, including assessment of tenders and awarding of contracts.
2. Communication is often also regarded as bilateral or multilateral, meaning that information is transferred back and forth between multiple parties. In this research, however, there is focus on what the public actors can do, in order to manage risk. One of these acts is providing information, here referred to as communication. Moreover, this action is recognizable in the situation of a procurement process in which a public organization often provides information on the project.
3. Personal communication, Dr J.C. Frost.
4. The ProCure21+ National Framework is a framework agreement with six supply chains selected via an *OJEU* tender process for capital investment construction schemes across England up to 2016.

REFERENCES

Bossink, B.A.G. (2007). The interorganizational innovation processes of sustainable building: a Dutch case of joint building innovation in sustainability. *Building and Environment*, 42(12), 4086–4092.
Chapman, C. and Ward, S. (2003). Project risk management: processes, techniques and insights.
Coase, R.H. (1960). The problem of social cost. *Journal of Law and Economics*, 3, 1–44.
Cowen, T. and Parker, D. (1997). Markets in the firm: a market-process approach to management. Hobart Paper No. 134. London: Institute of Economic Affairs.
Department for Business, Innovation and Skills (2009). Case study: forward commitment procurement HM Prison Service zero waste prison mattress system.
Department for Business, Innovation and Skills (2011). Forward commitment procurement – practical pathways to buying innovative solutions: Department for Business, Innovation & Skills.
Dorée, A.G. (1996). *Gemeentelijk Aanbesteden; een onderzoek naar de samenwerking tussen diensten Gemeentewerken en aannemers in de Grond-, Weg- en Waterbouwsector*. Proefschrift, Universiteit Twente, Enschede.
DTI (2006). Environmental innovation; bridging the gap between environmental necessity and economic opportunity.
Edler, J., Rufland, S., Hafner, S., Rigby, J., Georghiou, G., Hommen, L., Rolfstam, M., Edquist, C., Tsipouri, L. and Papadakou, M. (2006). Innovation and public procurement. Review of issues at stake. Study for the European Commission (ENTR/03/24). Fraunhofer: Fraunhofer Institute Systems and Innovation Research, Karlsruhe.
Edler, J. and Georghiou, L. (2007). Public procurement and innovation – resurrecting the demand side. *Research Policy*, 36(7), 949–963.
Edquist, C. and Hommen, L. (1999). Systems of innovation: theory and policy for the demand side. *Technology in Society*, 21(1), 63–79.
Edquist, C. and Hommen, L. (2000). Public technology procurement and innovation theory. In Edquist, C., Hommen, L. and Tsipouri, L. (eds), *Public Technology Procurement and Innovation*. Norwell, MA: Kluwer Academic Publishers.
Edquist, C. and Zabala-Iturriagagoitia, J.M. (2012). Public procurement for innovation as mission-oriented innovation policy. *Research Policy*, 41(10), 1757–1769.
Eisenhardt, K.M. (1989). Agency theory: an assessment and review. *Academy of Management Review*, 14(1), 57–74.
European Commission (2007). Guide on dealing with innovative solutions in public procurement – 10 elements of good practice. Luxembourg.
European Commission (2011). Public procurement – public purchasers

as launching customers. http://ec.europa.eu/enterprise/policies/innovation/policy/public-procurement/index_en.htm, accessed 16 December 2011.
Hommen, L. and Rolfstam, M. (2009). Public procurement and innovation: towards a taxonomy. *Journal of Public Procurement*, 9(1), 17–56.
LCB-Healthcare (2011). Workshop report: creating the conditions for innovation: towards a good practice guide.
LCB-Healthcare (2012). Innovation procurement: delivering efficiency, quality and sustainability in healthcare.
Lund Declaration (2009). The Lund declaration – addendum. Paper presented at the Swedish Presidency Research Conference in Lund: New Worlds – New Solutions Lund.
Makkonen, H., Aarikka-Stenroos, L. and Olkkonen, R. (2012). Narrative approach in business network process research – Implications for theory and methodology. *Industrial Marketing Management*, 41(2), 287–299.
Myoken, Y. (2010). Demand-orientated policy on leading-edge industry and technology: public procurement for innovation. *International Journal of Technology Management*, 49(1), 196–219.
Parker, D. and Hartley, K. (2003). Transaction costs, relational contracting and public private partnerships: a case study of UK defence. *Journal of Purchasing and Supply Management*, 9(3), 97–108.
Rutten, M.E.J., Dorée, A.G. and Halman, I.M. (2009). Innovation and interorganizational cooperation: a synthesis of literature. *Construction Innovation*, 9(3), 285–297.
Schotanus, F. (2005). Cooperative purchasing within the United Nations. Paper presented at the IPSERA 2005 Conference, Archamps, France.
Schotanus, F. (2007). Horizontal cooperative purchasing. PhD thesis, University of Twente, Enschede.
Tysseland, B.E. (2008). Life cycle cost based procurement decisions: a case study of Norwegian Defence Procurement projects. *International Journal of Project Management*, 26(4), 366–375.
Van Meerveld, H. (2012). The effects of the forward commitment procurement method on perceived risks in projects concerning public procurement of innovation. Masters thesis, University of Twente, Enschede, the Netherlands.
Van Meerveld, H., Nauta, N.J., Voordijk, J.T. and Whyles, G. (2012). The effects of the Forward Commitment Procurement method on perceived risks in procurement project concerning Public Procurement of Innovation in the healthcare sector. Paper presented at the HaCIRIC International Conference 2012, Transforming Healthcare Infrastructure and Services in an Age of Austerity, 19–21 September, Cardiff.
Von Hippel, E. (1986). Lead users: a source of novel product concepts. *Management Science*, 32(7), 791–805.

Ward, S. and Chapman, C. (2003). Transforming project risk management into project uncertainty management. *International Journal of Project Management*, 21(2), 97–105.
Williamson, O.E. (1979). Transaction-cost economics: the governance of contractual relations. *Journal of Law and Economics*, 22(2), 233–261.
Williamson, O.E. (1981). The economics of organizations: the transaction cost approach. *American Journal of Sociology*, 87(3), 548–577.
Winch, G.M. (2002). *Managing Construction Projects: An Information Processing Approach*. Oxford: Blackwell Science.
Yin, R.J. (2009). *Case Study Research: Design and Methods*, 4th edn. Thousand Oaks, CA: Sage Publications.

PART II

Case studies

6. Innovation and public procurement in the United States

Nicholas S. Vonortas

INTRODUCTION

The public sector accounts for a significant part of economic activity in Organisation for Economic Co-operation and Development (OECD) member countries. The average share of government expenditures in gross domestic product (GDP) is above 40 percent, varying from over 50 percent in Sweden and France,[1] to about 35 percent in Japan and the United States, to around 20 percent in Mexico (OECD, 2009). The performance of the public sector, along with its efficiency in providing public services, has been a major concern in democratic societies since ancient times. The primary source of this concern is the relative size of the public sector and the fact that it administers taxpayer funds. Pressures on the public sector for increased efficiency and productivity of resource use have mounted in recent years resulting in many calls for innovation, both process and product and service innovation.

As far as the public sector is concerned, process innovation means providing more (perhaps marginally improved) services at the same (or even lower) cost. Product innovation means providing new or significantly improved services. In either case, public expenditures for acquisition can influence the ability of the rest of the economy to produce new and significantly improved products and services, or to produce them more efficiently. This chapter focuses on this potential impact of public expenditure on innovation in the United States. I address public practices that induce innovation either: (1) by specifying levels of performance or functionality that are not achievable with off-the-shelf solutions; or (2) by supporting research and development (R&D) for goods and services to be procured afterwards, thus sharing the risks of the R&D

needed to develop innovative solutions. Following Edquist and Zabala-Iturriagagoitia (2012), I will call these practices public procurement for innovation (PPI). Pre-competitive procurement, whereby a government agency supports R&D without intention to purchase (Edquist and Zabala-Iturriagagoitia, forthcoming), is therefore excluded.

PPI has been defined as when a public agency acts to purchase, or place an order for, a product – service, good or system – that does not yet exist, but which could probably be developed within a reasonable period of time, based on additional or new innovative work by the organization(s) undertaking to produce, supply and sell the product being purchased (Edquist et al., 2000). Public procurement has been said to promote innovation in various ways: by creating new markets, by creating demand-pull, and by providing a testing ground for innovative products (Rothwell, 1984). Regarding market effects, in particular, public procurement can have three main roles (Fraunhofer, 2005): (1) market creation, where a market for the technology being procured does not yet exist; (2) market escalation, when an established market for the technology already exists but requires further development to succeed commercially; and (3) market consolidation, when standardized criteria or technical specifications for the technology to be used in the public sector lead to similar developments in the private sector.

PPI is currently being discussed worldwide in connection to ways through which governments can assist the development of new technologies and can promote businesses, research and, by extension, the economy (Edler and Georghiou, 2007). Though this issue has received much attention in Europe (Edler, 2013), it is hard to find research papers on US innovative procurement addressing cases where public procurement may have promoted innovation outside the national defense and security areas. It has been reported that although the United States has a strategic orientation in their public procurement, it is not connected to innovation per se (Fraunhofer, 2005). That study reported that most of the 'strategic' procurement in United States is geared towards achieving a social purpose like environmental protection, energy conservation and assisting disadvantaged groups.

This chapter deals with PPI in the United States outside the defense and national security areas. We find that the most important principles of the core federal procurement rules are cost savings and competitive processes. Innovation or priority for innovative products is not promoted as an essential policy consideration. Other important government regulations for procurement promote economy, effectiveness and efficiency. The substantive policies, however, are mostly left to the discretion of the acquiring federal agency, which is responsible for defining specifications and evaluating the criteria per procurement. For larger, more diffuse agencies this empowerment is frequently transferred down to the specific procuring bureau. In agreement with Fraunhofer (2005), I find that most of the 'strategic' procurement in the United States beyond defense and security is geared towards achieving some social objective. Innovation is occasionally the outcome of the quest for such social objectives.

The emphasis of procurement reforms in the past couple of decades on cost rationalization and minimization may run counter to investing in innovation, which is a costly and risky enterprise. Discouragement of innovation may be reinforced by tightening financial management, which adds administrative burdens possibly prohibiting the most innovative (small) organizations from participation. Chief acquisition officers report tendencies to overspecify procured product characteristics, mainly out of fear of litigation of unfairness claims. This runs counter to innovation where one would want to specify the desired performance characteristics and leave product characteristics to be proposed by the supplier. Still, one can find numerous examples of PPI by federal agencies. Their frequency relates to the mission and the general innovative culture of the agency.

The rest of the chapter runs as follows. The second section provides the main features of federal acquisition guidelines and delineates the overall trend in PPI. The third section presents the cases of five non-defense federal agencies that account for a large share of federal procurement, including the Department of Energy, Department of Education, Department of Health and Human Services, the Environment Protection Agency and the Department of Interior. The fourth section refers to the largest implementation challenges affecting public procurement and innovation. The fifth section illustrates important examples of public procurement inducing or leading to innovations by the Census Bureau, the Federal Aviation Administration, the National Oceanic and Atmospheric Administration and the Department of Energy. The final section concludes the chapter and draws lessons for the future.

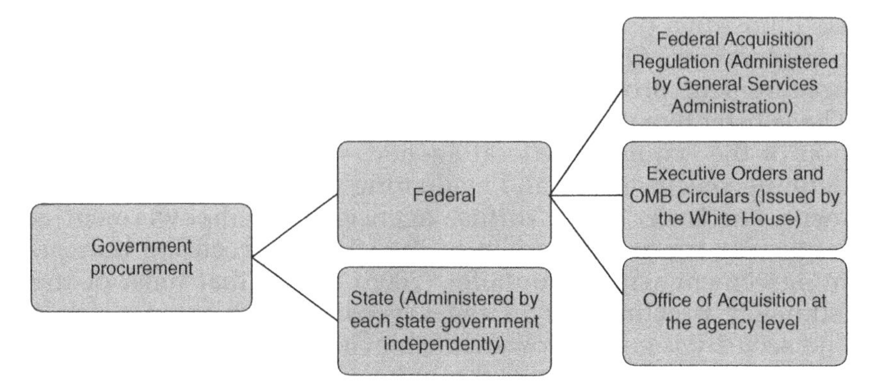

Figure 6.1 Public procurement matrix in the US

THE LIE OF THE LAND

The United States federal government is the world's largest purchaser of goods and services. It has been estimated that it paid out more than $700 billion (roughly 4.7 percent of GDP)[2] to contractors in the fiscal year 2011. This very significant public expenditure could, in principle, be better harnessed to promote innovation and, through it, strengthen the international competitiveness of the country.

Public procurement is carried out at two levels, the federal level and the state level (Figure 6.1). Though the bigger share of procurement is done by federal agencies like the Department of Defense and the Department of Energy, state governments have independent procedures set down for state-level procurement.

In the remainder I will concentrate on federal acquisition practice.[3]

Federal Procurement

Federal procurement is administered at three levels. The regulation binding on all federal agencies is the Federal Acquisition Regulation (FAR). FAR provides procedural rules that control the procurement process. Many federal agencies adapt FAR to their own specific needs; for instance, AIDAR for US Agency for International Development and HHSAR for Health and Human Services Department. FAR is generally considered the 'Bible' of federal procurement and is the binding authority for all federal agencies. Besides FAR, other important controlling documents are the Office of Management and Budget (OMB), circulars and memorandums and the executive orders from the White House issued from time to time as instructions to federal agencies. The third level of administration takes place at the agency level. Each federal agency has an in-house Acquisition Office that puts in place policies and procedures for that agency under the purview of FAR.

Federal Acquisition Regulation (FAR) and the General Service Administration

FAR lays down the procedures and general policies to be followed by the federal procurement. From time to time the General Services Administration (GSA) issues federal acquisition circulars to clarify changes in FAR. The two most important principles in FAR, considered inherent parts of all procurement decisions, are cost savings and competitive processes. Innovation or priority for innovative products is not promoted by FAR as an essential policy consideration.

There are five sections of the FAR most relevant to PPI:

1. FAR 9.202 states the process of qualifying a vendor for federal procurement, but it does not lay down the aspects that should be considered while selecting a vendor. Discretion is given to the individual agency to decide what they think should be a qualifying condition for a vendor. This clause reflects the powers given to each agency in deciding what is considered worth procuring and from whom.
2. FAR 11.002 lays down the policy to be followed while defining the needs of the agency. It provides that full and open competition be the underlying practice. Further, it provides that restrictions on the item specifications should be placed in very special circumstances where such restrictions are essential to achieve agency needs. This clause thus shows a similar trend of letting the agencies decide what is important for them. An interesting observation is made in Subpart 11.002 (2), where the agencies are actually required to define their needs in such a way that they can be met through 'commercial' or 'non-developmental' items.
3. FAR Subchapter D Part 23.2 states that federal procurement must take into account energy efficiency and environment-friendliness. Still, cost efficiency is provided as an innate

principle for determining the worthiness of a product or process. Though FAR does not describe this policy and its rules in detail here, this aspect of US government policy at the agency and sector level is discussed in detail below.

4. FAR Part 35 separates acquisitions with the sole purpose of R&D from acquisitions for direct benefit of the government. Nonetheless, the same rules of acquisition apply to both types of federal procurement.
5. FAR Part 48 mentions value engineering as the main evaluation method. This method is used to apply 'best value' as the main evaluation criterion.

Executive orders, OMB circulars and memos

The Office of the President issues executive orders as binding instructions for federal agencies. The Office of Management and Budget (OMB) at the White House similarly issues circulars and memorandums on administrative matters for federal agencies. These documents consist of procedural details along with policy considerations fueling the instructions. They typically aim at promoting economy, effectiveness and efficiency. The OMB's Office of Federal Procurement Policy (OFPP) is responsible for issuing such procedural guidance for federal agencies regarding procurement matters. These orders, circulars and memos, though not the rule of law, become procedural guidelines for federal agencies and thus impact the application of rules laid down by the Congress.

For the purpose of this chapter, I have looked at all executive orders, circulars, policy letters, guides and memos, released by the Office of the President and the OFPP of the White House during the period 2006–2010, that related to federal procurement, in order to establish possible relationships between innovation and procurement policies. In a nutshell, the White House documents point to the fact that the government's focus on innovative solutions is generally aimed at achieving some other social purpose. More specifically, three social purposes were identified to link procurement and innovation: development of accessible technologies by disadvantaged groups, the improvement of space technologies, and the development of environmentally friendly products. The government encourages the development by contractors and procurement of innovative solutions to meet these three types of social needs. Direct reference to developing better technologies through federal procurement was found in a single executive order (13423). In this order, among many other goals, federal agencies were given the goal to acquire sustainable, environmentally friendly and green products.

Agency-level acquisition guidance

More than 50 federal agencies are involved in activities that require acquisition of goods and services. Each agency has a fair amount of discretion in setting its standards and procedures under the wide umbrella of process rules laid down by FAR. FAR Subpart 1.4 provides for issuing regulations and policies that may deviate from FAR if so required by the agency.

Most agencies have an in-house Office of Acquisition that provides guidance on procurement issues by interpretation of FAR, executive orders and OMB circulars and memos. Other important roles played by the Acquisition Office include approval of specifications for procurement and oversight of the competitive bidding process. The discretion to state the actual specification for the procurement remains largely with the technical heads of the agency placing the order.

While procurement in the United States is subject to a mesh of regulations issued and administered by a wide range of actors, significant decision-making power is left to the procurement manager in each agency.[4] The substantive policies are mostly left to the discretion of each acquiring federal agency, which defines both the specifications and the evaluating criteria per procurement.

The independence of each bureau to come up with its procurement needs is based on the overall federal practice of giving maximum discretion to the end client or to the acquiring agency in laying down procurement rules fitting to the agency's overall mission. The only restrictions laid down on procurements across the nation include the cost-effectiveness and open competition requirements of the FAR, along with environmental considerations through vehicles like the executive orders, OMB memos and guidance from the Environmental Protection Agency under different regulations.

FEDERAL PROCUREMENT: ACQUISITION MECHANISMS IN FIVE AREAS

I present the cases of five non-defense federal agencies – the Environment Protection Agency, the Department of Energy, the

Department of Health and Human Services, the Department of Education and the Department of Interior – that account for a fairly large share of federal procurement.

Environment

The environment has been one of the biggest concerns for the country for a long time. This has given the Environmental Protection Agency (EPA) a special place among federal agencies. The agency has been given numerous powers to regulate and administer laws and regulations that affect the environment.

The tendency to develop innovations while working towards some social cause is more visible in the environment sector than anywhere else. Concerns for clean air, water and other natural resources, along with the need for alternative clean sources of energy, underline a lot of activity in research and innovation. Innovation is promoted in all fields including energy sources, construction material and even cleaning supplies in order to develop environmentally friendly products and processes.

EPA conducts its own procurement through the Office of Acquisition Management which, like all other agencies, has developed its own version of FAR, called the EPA Acquisition Regulations (EPAAR). The agency follows the general principles of 'full and open competition' and cost-effectiveness in all its procurements. However, in addition to following the policies laid down by FAR, EPA is also responsible for laying down policy and issuing guidance for all federal agencies under Title 40 of the Code of Federal Regulations, the Clean Air Act, Clean Water Act, Energy Policy Act and many more.[5]

In order to administer the numerous relevant regulations and executive orders, EPA provides guidance in all fields including clean water, clean air and clean energy. Most of these regulations are aimed at the private sector. EPA plays various roles:

1. Lawmaking: EPA defines and sets regulatory standards for environmental practices on the basis of the laws and orders provided by the government. EPA also adds regulations and limitations for other federal agencies to follow.
2. Policing: EPA is responsible for checking that rules laid down by the law are followed by both the private and the public sectors.

3. Resource library: EPA provides numerous resources to industries and other public agencies in need of information and guidance for making their facilities more environmentally friendly. It is in this role that EPA becomes an important partner for other federal agencies in meeting the requirements placed on them.

One of the most important functions of EPA is to provide the basis for focus on PPI in the environment area. Public demand for innovative products in turn sends strong signals to private users. Thus developmental catalytic procurement becomes the heart of procurement with the aim of market creation, escalation and consolidation.

Public procurement in the environmental area is one of the best examples of PPI in the United States. Federal agencies are asked to do their part through procurement. While innovation is not the underlying social cause, it is frequently the end result.

Energy

Energy attracts a lot of policy attention regarding the need for innovation, especially in relation to competitiveness and energy independence. Energy independence and development of alternative energy technologies have been a national concern for several decades, thereby making innovations in the field of energy a national priority.[6] The Department of Energy (DoE) spends a huge amount (90 percent) of its budget on a variety of contracts and federal assistance agreements. This makes the agency unique. The basic mission and intra-agency culture of DoE place innovation at center stage.

DoE is subject to FAR for general rules on procurement. However, for its specific acquisition purposes, DoE has developed the Department of Energy Acquisition Regulations (DEAR). DEAR is formatted to incorporate most of the FAR rules and has a similar order of the procedural regulations. Part 935, as in FAR, refers to additional procedural requirements for research and development contracting. However, the only purpose of this regulation is to lay down additional reporting requirements for contracting done for research purposes. The discretion of deciding what to procure and with what specifications remains with the agency.

As in the case of the environment, developmental public

procurement is the main form of PPI in the energy sector. The aim is to use cooperative and catalytic forms of public procurement for market creation, escalation and consolidation.[7] Purchasing by state and other public actors in this sector is directed not only towards fulfilling their own original missions, but also towards influencing and supporting certain patterns of demand on the part of private consumers.

The energy sector invests a lot on innovation and research. The DoE goes a step further and provides its contractors with direct support to commercialize these technologies. For instance, DoE's website provides names and links to all its contractors' websites and encourages visitors to contact these contractors for business opportunities. Such a feature provides exposure to new technologies and helps in integrating them into the economy.

Healthcare

Public health has been one of the expressed core missions of US science policy for several decades (along with national defense and the support of basic research). Healthcare and healthcare delivery have also been at the forefront of policy concerns in the country as a result of the major health reform passed by President Obama. In the public sector, focus seems to be more on healthcare delivery. Innovation, then, is mainly focused on healthcare delivery aimed at improving the poor state of healthcare services for many Americans.

Not much innovation is being fueled through procurement. Funding programs have, of course, an effect on the technologies used in the field and may thus qualify as public money resulting in innovation. The healthcare reform has many such small programs that affect healthcare delivery in many ways. Though it is not possible to discuss at length each program and its effects on innovation, the example of a $100 million program launched by the Department of Health and Human Services Secretary helps explain the picture. This program, launched in 2009, gives money to ten states to test innovative technologies in healthcare delivery for children.

Such programs, though not strictly procurement, help to promote innovation by focusing on social causes in need of innovative ideas and technologies. However, as they are not administered in the form of procurement, the regulations affecting their administration are very different.

The Department of Health and Human Services (DHHS) is the relevant federal agency in this area. DHHS – through its National Institutes of Health (NIH) – is the largest public funder of research in the country after the Department of Defense. However, according to the definitions used in this chapter, the vast majority of this funding does not qualify as public procurement and is excluded from this discussion.

Procurement is administered at DHHS through the Office of Grants and Acquisition Policy and Accountability's Division of Acquisition. Like other federal agencies, DHHS has also developed a custom version of FAR known as the Health and Human Services Acquisition Regulation (HHSAR), which provides specific procedural rules to be followed during procurement. Again, no specific policies or guidelines are provided to include specific criteria in the scope of work. The scope of work and its content is to be based on the needs of the agency that are being met through procurement with the overall FAR principles of 'best value' being implemented through value engineering.

Infrastructure

The Department of Interior (DoI) is the federal agency responsible for the development and maintenance of public infrastructure. DoI has many bureaus that manage specific fields like land management, ocean energy management, fish and wildlife, surface mining, and so forth. Each bureau plays a specific role and conducts its procurements under the guidance of the overall acquisition management system of DoI.

Procurement is administered by the Office of Acquisition of Property Management (PAM) as per the tailored version of FAR called Interior Department Acquisition Regulations (IDAR). Procurement requirements are controlled to a very high degree by the acquiring agency. While PAM is the acquisition intermediary for all bureaus of DoI, each bureau independently defines its needs and publishes its own list of procurements on the websites.[8]

There is a vocal recognition in the US government of the deteriorating condition of the nation's infrastructure and a push to prioritize its development. The application of new technologies is seen as a consequence. DoI regularly spends one-fourth of its budget on small contractors for regular infrastructure maintenance-based assignments. In addition, the 2009 American

Recovery and Reinvestment Act gave the Department of Interior extra money to invest in infrastructure. The main objective of that one-time funding was job generation and fighting unemployment. Though DoI spends significantly on infrastructure, most of its investments come across as regular everyday maintenance jobs rather than new technology development. The agency does not indicate the same level of enthusiasm as in the energy and the environment fields where innovation frequently appears to be at the heart of activity.

Education

Education is generally managed at the state level. However, the federal Department of Education (DoEd) is responsible for setting standards and rules that must be followed by the states in order to receive federal funding for education. DoEd also invests in developing better and improved learning methodologies.

DoEd's Office of Chief Financial Officer, through the Contracts and Acquisition Management group, is responsible for procurement. DoEd has also tailored FAR for its purposes into the Department of Education Acquisition Regulation (EDAR). The regulation is similar in structure and content to FAR, and simply modifies the words and definitions to suit its needs while maintaining the rule itself. No special mention or interest has been given to innovation. The decision on whether to make innovation a criterion in procurement rests on the department requesting the acquisition.

DoEd launched the grant program entitled Investing in Innovation (I3) under the American Recovery and Reinvestment Act of 2009, with a focus on developing innovative techniques to improve school performance. Though this program was executed by the Office of Innovation and Improvement and was designed as a grant, it provided the much-needed insight into the US government's policies regarding promoting innovation for improving everyday lives. Through this program, the DoEd requested assistance from schools, local-level educational bodies and non-profit organizations to use their proven innovative techniques in education to develop innovative programs and strategies. The total amount of the program was $650 million. The main objective was to improve school-level performance for children. Innovation was promoted indirectly under this main objective. While this program is directly relevant to my study, it was shaped as a grant, and thus does not qualify as procurement-backed innovation. Different rules apply to procurement and grants.

Concluding this section, one observes important differences among various areas in terms of promoting innovation through public procurement. In all cases, however, innovation is seen as a means to achieve other worthy social objectives. It is these social objectives that push agencies to innovate and to promote new technologies.

REAL CHALLENGES IN REAL TIME

Chief acquisition officers in federal agencies do speak about innovation, but about innovations in the process of procurement. Almost every acquisition officer that I and my colleagues who assisted during the case study (see Acknowledgments) spoke with[9] is evaluating and thinking about how to make the procurement process more efficient, to create cost savings, and to better align the process of their requests for proposals and the end products that are received. These are obviously not new topics, but it is clear that the current US Administration has taken an interest in this area and has been pushing for changes. The White House Office of Federal Procurement Policy (OFPP) has created committees with large and small agencies to discuss what they refer to as innovative procurement practices. The common topics of interest are leveraging the purchasing power of the federal government, creating a cloud computing system, training the incoming generation of acquisition officers, and learning from business to understand how to have the government get what it actually wants.

A key factor differentiating between the success and failure of an acquisition is in the writing of the requirements. More than anything else, this area seems to provide the connection between the procurement tools and the acquirer's needs. There is a major challenge in the procurement process, however, which has to do with fairness and the fear of the litigation that may ensue after the selection of an awardee. To safeguard against litigation, most procurement officers have a tendency to overspecify the requirements of the product or process that they need, rather than allowing those in the field to bring the latest technology to the table. Overspecification stifles innovation.

We were able to isolate only a few examples where the process of what we call PPI had worked well: the delivered product or system had never been created before, was state-of-the-art, and met the needs of the agency. When asked specifically to draw upon the lessons learned from those examples, the answers of procurement officers were a bit too obvious and a bit unnerving for those looking for systemic approaches. The actual process for procurement appears to be fairly similar across the agencies. That is not to say their content is the same, but the processes have a number of noticeable similarities. A procuring activity typically arises from a need within the agency. A program director alerts the procurement office that they have the budget to put out a solicitation to fill that need. The program director or the acquisition officer, or frequently these two together, write the specifications for the solicitation. It is unclear what the background of either individual is when writing the specifications. There is no well-specified process in place from the procurement perspective to ensure that the person writing the specifications truly understands upfront what they need. Going back to the successful cases, each time the success was attributed largely to a specific person making sure the procurement ran smoothly. This person usually reached out to businesses and sought advice upfront, then made certain the project remained a priority, and worked closely with the contractor to make sure it was offering what was necessary. From both the business perspective and the federal perspective, the communication between the parties, upfront and during the process, is really key to success.

It is evident that the procurement officers place heavy emphasis on the tools rather than the content of the purchase. As long as the protocol is met, the program directors have significant freedom in what they purchase. Each government agency approaches the subject of innovation differently. If an agency's mission includes research, it appears that when they create a solicitation, they are more willing to describe their needs and allow proposers to suggest the method for reaching those goals. Further, such agencies include innovative strategies in the reviewer requirements. This is certainly the case of the National Aeronautics and Space Administration (NASA) and NIH. Most of the procurements from both agencies are left fairly open. However, as it is the job of the procurement officer to make sure the solicitation and the award are successful, it takes the senior leadership of the specific agency to dictate the tone.

If the leadership's focus is around cost savings, or minimizing the number of appeals, then the procurement office will be less willing to allow broader solicitations.

A transformation has begun in the United States over the past few years. Through the OFPP working groups, a community of procurement officers has formed. Many realize they face similar issues and that they are not taking full advantage of the flexibility FAR provides and the power at their fingertips. There will likely be a more strategic push in the near future from this group to change the procurement process in the direction of fostering innovation more proactively.

INSTANCES OF PUBLIC PROCUREMENT FOR INNOVATION

This section will introduce three examples where PPI has been at the core of the public intervention: the TIGER system, the ADS-B program and the NOAA's remotely operated vehicles and high-performance computing.

Census Bureau: TIGER System

Background

The Census Bureau serves as the leading source of quality data about the US nation's people and economy. The Bureau conducts the US Census, which is mandated by Article I, Section 2 of the Constitution, and takes place every ten years.

Prior to the 1960 Census, enumerators of the Census Bureau visited each household and asked the residents to fill out a questionnaire. In order to avoid duplications and speed up the process, the Bureau provided enumerators with maps, showing the assigned area. In the late 1960s, the Bureau developed the Dual Independent Map Encoding (DIME), a geographic information system to handle spatial data (Tomlinson, 1990). Despite successful implementation of the DIME files of major metropolitan areas, many problems were encountered during the 1980 census due to the fact that three basic components of geographic support – a national address list, geographic reference files and maps – were not coordinated, which resulted in numerous problems during field operations, tabulation and data dissemination. In order to reduce the likelihood of future problems and ascertain consistency among and between geographic support functions and products, the Bureau embarked on an ambitious project to create a national digital spatial database to support the next two decennial censuses.

TIGER system

At the August 1984 11th International Cartographic Conference, Robert W. Marx, then chief of the Geography Division at the US Census Bureau, presented a paper describing the agency's concept for a seamless nationwide geographic database that would facilitate the collection, processing, dissemination and understanding of statistical data about the people and economy of the United States (Marx, 1984). The new database would contain information about all the physical features commonly found on the maps used by the Bureau's enumeration staff (for example, roads, railroads and selected landmarks), the street name and address range information required to automatically geocode most city-style addresses and associate each with the road along which it was located, and all the geographic boundaries, codes and names for the governmental units and statistical entities used for interviewing, data collection and delivery of the tabulated statistical data (Tomasi, 1990).

Although the concept seemed radical and unlikely to succeed, the Geography Division broke new ground by developing a valuable national resource to assist the census, the Topologically Integrated Geographic Encoding and Reference System (TIGER) geographic database. TIGER is the system and digital database that supports the decennial census and other Census Bureau statistical programs. The topological structure of the TIGER database defines the types, locations, names and relationships of streets, rivers, railroads and other geographic features to each other and to the numerous geographic entities for which the Census Bureau tabulates data from its censuses and household surveys. TIGER expanded earlier efforts in digital spatial database development and emerged in a fundamentally operational state in only six short years. It was viewed not only as a data repository but as a 'system' which included the data, the database, all software applications, documents and procedures. The TIGER database does not contain demographic data but merely map data such as street address ranges, zip codes and feature names available to the public free of cost.

TIGER has been used extensively within the Bureau. The

database and its suite of applications software have been used to provide the geographic framework and maps required to complete several censuses of the United States, Puerto Rico and Island Areas, several economic censuses, numerous special censuses and census tests, monthly household surveys, and contributed to the development of a continuous measurement system known as the American Community Survey.

Additionally, TIGER has been used by local and tribal governments and by private industry for other purposes. This can be largely attributed to the open access policy of the United States federal government.[10] This practice has been the fuel in the explosion of geographic information system (GIS) technology and relevant applications throughout all levels of government, academia and much of the private sector in the United States. The public products of the TIGER database allowed the then nascent GIS industry of the United States to flourish by devoting energy to the development of powerful and innovative display and analysis tools rather than to gathering, updating and disseminating geographic base data. Instead of competing with the private sector developers of GIS software products, the Census Bureau works actively to assure that all interested parties know what the next planned formats of the TIGER extracts will be, so that each company can prepare the tools and packages it believes to have the greatest commercial value.

TIGER was developed and procured to support and improve the Bureau's process of taking the Decennial Census. It ended up as an innovative tool for the development of a whole industry, GIS, which was largely based in this technology in order to flourish.

Federal Aviation Administration: Automatic Dependent Surveillance-Broadcast (ADS-B) Program

ADS-B is one of the most important, underlying technologies in the Federal Aviation Administration (FAA)'s Next Generation Air Transportation System (NextGen) to transform air traffic control from the current radar-based system to a satellite-based system. NextGen focuses on leveraging new technologies, such as satellite-based navigation, surveillance and networking, and involves collaboration among government and the private sector in aerospace and related industries. Essentially it is a technology solution that pinpoints an aircraft's location using satellite

Global Positioning System (GPS) navigation, and allows the aircraft to constantly broadcast its precise location and other flight data (for example, altitude, velocity) to nearby aircraft and air traffic controllers. ADS-B will for the first time allow both pilots and controllers to see the same real-time displays of air traffic, thus improving safety and air traffic management.

The FAA went through a multi-stage contracting process to ensure the selection of the best vendor for this project. Prior to issuing a request for offers (RFO), the agency held multiple 'industry days' that allowed interested businesses to present their thoughts on the development of the ADS-B system. These presentations then directly influenced the writing of the contract requirements. Following these meetings, the FAA requested a screening information request (SIR), which allowed the serious competitors to come forward and submit a proposal. Teams headed by ITT Corp., Lockheed Martin and Raytheon were found to have viable solutions. The agency then issued an RFO in March 2007 that officially asked the three vendors to submit their proposals for providing ADS-B services. A team of subject-matter experts in technical, business and cost areas evaluated and scored each proposal, based on strict evaluation criteria. Based on this extensive analysis, the FAA in August 2007 awarded the contract to the team headed by ITT Corp. because its proposal combined the best value and presented the least risk for a successful implementation.[11]

ADS-B has been structured as a service contract. Vendors will install and maintain the ground-station equipment, and the FAA will pay subscription charges to the vendor, just as the agency today buys telecommunications services from telecommunications companies. This is known as a performance-based service acquisition. This service-based acquisition is possible in part because ADS-B ground stations are small and can be deployed nearly anywhere, unlike huge radars the government owns today. From the government's point of view, a performance-based service acquisition maximizes the competition and substantially reduces the costs incurred by leasing land and owning, maintaining and upgrading equipment throughout the system's lifecycle. It also allows vendors to use facilities and equipment they already own – such as cellphone towers and support buildings, fiber-optic networks and operations centers – so they can piggy-back off existing assets. As a result, actual hardware investments will be relatively small since much of the infrastructure is already in place and being used for other purposes. This speeds the deployment and reduces costs. Finally, by purchasing services instead of equipment, the FAA can easily and quickly adapt to local increases or decreases in air traffic volume that change the level of services required. Once the ADS-B infrastructure is in place, vendors will likely use the system's capabilities to offer even more services to pilots and airlines.

During the entire process, the FAA remained in direct communication with the vendors and the industry. In addition, the services contract allowed a lot more flexibility in the responses that the vendors could provide. A quote from the ADS-B National Airspace System Wide Acquisition Program Advance Notification to Industry issued in April 2006 really makes the agency's flexibility and openness known:

> It is believed that a performance specification will allow greater flexibility in execution of a solicitation. With a performance specification, the FAA will have an option of a 'traditional' solicitation wherein offerors would present designs, manufacture, test, and deliver equipment to [the] FAA or an option where offerors would propose their approach for satisfying the performance service requirements using their own facilities and equipment without any delivery to or ownership by [the] FAA. The latter option allows industry to utilize and maintain equipment that does not meet all the FAA imposed requirements such as wiring bend radius, physical dimensions, markings, and so on. Provision of a service would be the contract deliverable, in the same manner as cell phone and cable TV providers . . . Consideration is being given to a performance-based acquisition whereby industry would be given the opportunity to create contract pricing terms and conditions. The FAA is interested in unique approaches which allow both incentives to the contractor and penalties for non-performance. [The] FAA is open to discussion of methodologies to reduce the overall cost of broadcast services through potential added opportunities created for broadcast services providers. The BSGS [Broadcast Services Ground System] draft specification, modified as appropriate by industry comment, will also be provided as guidance and information, possibly before any formal solicitation.

This quote represents why the contract has seemingly been successful. The level of communication, direct connection to management, and the allowance of business to bring their ideas to the forefront have fostered an open and innovative environment.

The contract award to ITT Corporation was for $1.8 billion (if all options are exercised through 2025) to develop and deploy the ADS-B ground infrastructure and start broadcasting services. FAA plans to implement ADS-B in two segments. Segment 1 (establish ground infrastructure for five key sites) is contracted under a cost-plus incentive fee agreement in which FAA covers the cost for any additional requirements. Segment 2 (equipment needed to fully deploy ADS-B nationwide) is contracted under a fixed-price arrangement in which ITT covers the cost of deploying enough radios to meet requirements. The FAA approved nearly $1.7 billion in capital costs through 2014, to support ADS-B implementation. The total life-cycle cost through 2035 of the ADS-B effort is uncertain but estimated to be about $4.0 billion, including $2.3 billion in capital costs not yet formally baselined.

ADS-B uses GPS signals along with aircraft avionics to transmit the aircraft's location to ground receivers. The ground receivers then transmit that information to controller screens and cockpit displays on aircraft equipped with ADS-B avionics. According to FAA, ADS-B will supplement and ultimately replace ground-based radar because an ADS-B equipped aircraft can provide controllers and pilots in other aircraft with faster updates of important flight information (for example, aircraft identification, position, altitude, direction and speed). Specifically, ADS-B transmits position information once per second, whereas radar systems in the vicinity of airports generate reports once every 4–5 seconds. Also, unlike radar, the accuracy of ADS-B does not change based on the distance between the aircraft and the sensor. More specifically, the current Ground Radar-Based System and the ADS-B System compare as follows:

- Lower cost, higher accuracy and more frequent updating of information. ADS-B infrastructure consists of relatively simple radio stations which are cheaper to install and maintain than traditional radar that requires significant mechanical infrastructure and signal processing. ADS-B is also more accurate at identifying aircraft and determining position. The ADS-B System is updated by aircraft every second, compared to once every 12 seconds for en route radar systems. And ADS-B provides three-meter accuracy, which combines for increased operating efficiency in areas of dense traffic.

- Full airspace coverage. ADS-B equipment can be installed in areas where it is not feasible to establish radar-based surveillance equipment. For instance, ADS-B equipment will be installed on oil platforms in the Gulf of Mexico, delivering substantial safety and efficiency benefits to air traffic flying over the area.
- Improved cockpit safety. ADS-B equipped aircraft will be able to receive and display in the cockpit the position of all other ADS-B equipped aircraft in the area. Until all aircraft can be fully equipped, the Traffic Information Services Broadcasts (TIS-B) will provide situational awareness to ADS-B equipped aircraft by identifying radar targets of non-ADS-B equipped aircraft. FAA's ADS-B concept also provides Flight Information Service Broadcasts (FIS-B) to provide pilots with current weather information and awareness of meteorological conditions that might impact flight.
- Increased airspace capacity and efficiency. ADS-B provides a vehicle for increased cockpit involvement in the air traffic control process. This may represent the most significant potential benefit of ADS-B in terms of airspace capacity and efficiency. The ability of aircraft to have a cockpit display of all surrounding traffic enables air traffic control procedures that begin to involve the participation of the cockpit crew. Several applications of this capability have been defined and are in the process of achieving operational certification. For instance, Cockpit Based Merging and Spacing can allow a pilot to lock onto a preceding aircraft and to maintain a very precise spacing interval. This offers the ability to optimize the arrival spacing at busy terminal areas and to make maximum use of Continuous Descent Arrival (CDA) procedures delivering fuel savings and reduced emissions.

Table 6.1 compares ground radar and ADS-B systems. Figure 6.2 illustrates the life-cycle management process of ADS-B and critical decision points.

Table 6.1 A comparison between ground radar and Automatic Dependent Surveillance Broadcast (ADS-B) systems

Ground radar-based system	ADS-B system
On the ground, dependent on human participation	On the aircraft, providing a constant flow of more accurate identification and location data
Coverage gaps exist in some areas	ADS-B ground stations can be placed anywhere (for example, mountains, oil rigs)
Positions updated by aircraft every 12 seconds	Positions updated by aircraft every second
Costly to install and maintain	Significantly less costly to install and maintain

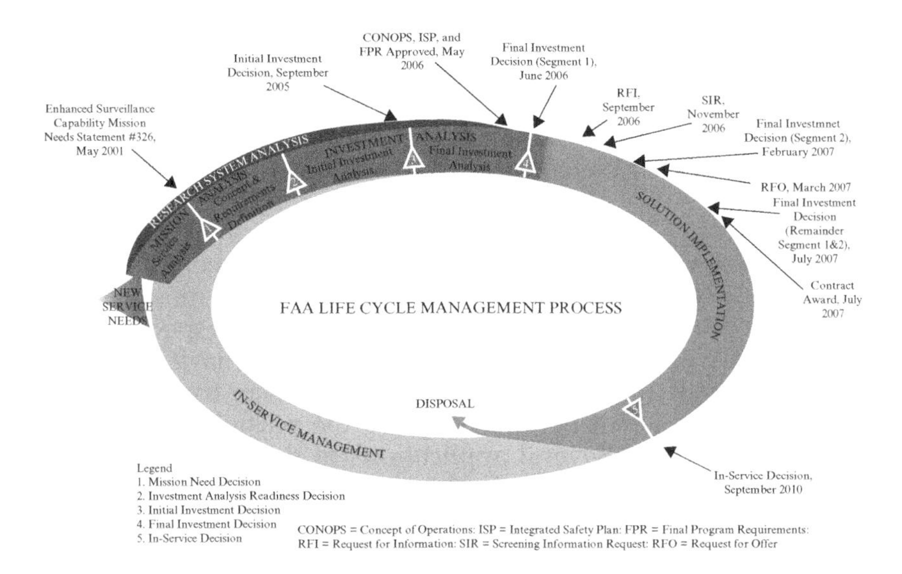

Source: Robert Strain, 'Surveillance and Broadcast Services', MITRE, presentation Weather in the Cockpit Workshop, 8 August 2006, www.ral.ucar.edu/projects/wic/references/wic1workshop/strain1.ppt. Accessed 24 November 2013.

Figure 6.2 FAA life cycle management process for ADS-B

National Oceanic and Atmospheric Administration: Remotely Operated Vehicles and High Performance Computing

Background

The National Oceanic and Atmospheric Administration (NOAA), a scientific agency within the United States Department of Commerce, focuses on the conditions of the oceans and the atmosphere. NOAA's primary mission is to warn of dangerous weather, chart seas and skies, guide the use and protection of ocean and coastal resources, and conduct research to improve understanding and stewardship of the environment. The agency conducts an end-to-end sequence of activities, beginning with scientific discovery and resulting in a number of critical environmental services and products relating to: monitoring and observing Earth systems with instruments and data collection networks; understanding and describing Earth systems through research and analysis of that data; assessing and predicting the changes of these systems over time; engaging, advising and providing the public and partner organizations with important information and managing resources for the betterment of society, economy and environment.

NOAA promotes the introduction of new technologies that will enable it to describe, understand and predict the environment. Core technologies include:

- Biodegradable sensors in target environments that are cheap enough to simply be replaced rather than maintained.
- Mobile sensor platforms (for example, unmanned aerial vehicles and remotely operated undersea vehicles) to monitor and observe the land, the atmosphere and the ocean surface, depths and floor.
- Information technology to advance model-based analysis techniques to describe, understand and predict the interactions of all parts of the environment at increasingly finer resolution.
- Telecommunications that will have the capacity to link modeling and ecological information centers seamlessly and effortlessly, thus providing users with the capability to 'reach back' to powerful high-performance computers, taking advantage of state-of-the-art modeling and forecasting, to meet their own, individual needs.

Example 1: remotely operated vehicles (ROV)

In the last 50 years technology has advanced to the point where humans can examine the oceans in systematic, scientific, and, most importantly, noninvasive ways. The National Ocean Service (NOS) line office is focused on ensuring that ocean and coastal areas are safe, healthy, and productive. Remotely operated vehicles (ROVs) are among the innovative technologies that make today's explorations possible, such as the numerous vessels, submersibles, diving technologies and observation tools that enable researchers to examine, record and analyze the oceans.

An ROV is an unoccupied, highly maneuverable underwater, tethered robot usually equipped with a video camera and lights. It is linked to the ship by an umbilical cable, a group of cables that carry electrical power, video and data signals back and forth between the operator and the vehicle. High-power applications will often use hydraulics in addition to electrical cabling. ROVs are highly customized and their equipment may include a still camera, a manipulator or cutting arm, water samplers, and instruments such as sonars and magnetometers that measure water clarity, light penetration, temperature and other parameters.

First developed in the 1960s with funding from the US Navy, most of the early ROV technology development resulted into what was then named a 'cable-controlled underwater recovery vehicle' (CURV). This created the capability to perform deep-sea rescue operation and recover objects from the ocean floor. Building on this technology base, the offshore oil and gas industry created the work-class ROVs to assist in the development of offshore oil fields. Since then, technological development in the ROV industry has accelerated and today ROVs perform numerous tasks in many fields. Their tasks range from simple inspection of subsea structures, pipelines and platforms, to connecting pipelines and placing underwater manifolds. They are used extensively in both the initial construction of a sub-sea development and the subsequent repair and maintenance.

ROVs represent an example of a technology that was developed for a purpose and served many others. Through its R&D funding over the years NOAA has largely contributed to this direction. While it is true that the oil and gas industry uses the majority of ROVs, other applications include salvage, science and military. ROVs are often kept aboard vessels mounting submersible operations for several reasons. They also support exploration and science objectives. If a submersible cannot be used because of weather or maintenance problems, the ROV can often take its place. Lastly, the military uses ROV for tasks such as mine clearing and inspection.[12]

Example 2: high-performance computing

Telecommunications and information technology are core technological competences of NOAA. The Office of the Chief Information Officer (OCIO) is responsible for ensuring that NOAA's programs make full and appropriate use of information technology (IT), and oversees the expenditure of approximately $600 million annually in IT hardware, software, services, networking and telecommunications.[13] As a huge buyer in high-performance computing, NOAA has established with its NAO 216–110 Order a policy for managing high-performance computing resources as a corporate asset in support of its mission.[14] The policy states that high-performance computing (HPC) will be managed by the NOAA environmental modeling program (EMP) for the benefit of NOAA as a whole.[15] HPC uses supercomputers and computer clusters to solve advanced computation problems. Computer systems approaching the teraflops capability are counted as HPC computers.

The objectives of HPC use are: (1) to enable the agency to disseminate its vast holdings of real-time and historical information to users more completely, in a more usable form, and in a much more timely manner via advanced networking; and (2) to enhance NOAA's scientific productivity through the use of advanced collaboration and model analysis tools, enabling faster, more effective communications among researchers, and improved analysis, diagnosis and visualization of model output. To meet this demand, NOAA scientists require leadership-class, high-performance computing (HPC) systems with petaflops-scale capabilities. NOAA currently has dozens of major models and hundreds of variants of models under development that support the full range of forecasting challenges that are maintained in the operational phase, producing multiple informational products daily for public use as well as for special needs, such as emergency management. New high-performance computing hardware architectures require scientific applications to run across multiple processors, rather than a single processor, to achieve desired performance. In order to accommodate this new large-scale supercomputing approach to modeling, NOAA has determined that it requires a new, more flexible HPC architecture. This target architecture must span a vast array of technical and product delivery requirements to meet the needs of millions of diverse stakeholders, including NOAA scientists, academic researchers, private sector planners, federal partners and the general public, particularly when life and property are threatened. In the case of NOAA, contractors are asked to push the envelope and give the most innovative solutions. Manufacturers are provided with a specific set of requirements that will allow the development of high-performance computing systems that will accommodate the specific demands of NOAA.[16]

Example 3: from Doppler radar to dual polarization radars
NOAA's National Weather Service relies daily on radar to detect, locate and measure precipitation inside clouds. Radar technology designed to detect and locate hostile aircraft and missiles in World War II serves as the basis for today's advanced weather radar systems. Radar is an object-detection system which uses electromagnetic waves to determine the range, altitude, direction or speed of both moving and fixed objects such as aircraft, ships, spacecraft, guided missiles, motor vehicles, weather formations and terrain. Due to its ability to provide information regarding the position of a variety of objects ranging from aircrafts to rain droplets, it can be used in many different fields where the need for such positioning is crucial. Examples are aviation radar, marine radar and even vessel traffic service radar systems whereby radar is used to monitor and regulate ship movements in busy waters and to monitor vehicle speeds on the roads.

Meteorologists use radar to monitor precipitation. It has become the primary tool for short-term weather forecasting and to watch for severe weather such as thunderstorms, tornadoes, winter storms, precipitation types, and so on. The WSR-88D weather radar system (Weather Surveillance Radar –1998 Doppler) was developed by the National Weather Service during the mid-1980s and fully deployed by the early 1990s. Doppler radars have the added capability to measure a frequency shift that is introduced into the reflected signal by the motion of the cloud and precipitation particles. This frequency shift is then used to determine wind speed. There are 158 such radar sites in the United States and selected overseas locations. The WSR-88D weather radar provides high-resolution data and the ability to detect intra-cloud motions, and has been the cornerstone of NWS severe weather warning operations. Data gathered by radar and related systems are processed by meteorologists using an advanced information processing, display and telecommunications system called AWIPS.

The entire fleet of WSR-88Ds was scheduled for a major software and hardware upgrade (by mid-2013). This upgrade, known as dual-polarization technology, will greatly enhance the radar by providing the ability to collect data on the horizontal and vertical properties of weather (for example, rain, hail) and non-weather (for example, insect, ground clutter) targets.[17] Dual-polarization capabilities were added to the Cimarron Doppler radar in time for the 1985 spring storm season. Scientists had learned that when pulses were alternately polarized vertically and horizontally, the return signal provided a clearer indication of cloud and precipitation particle size, shape and ice density, and they could determine whether the targets were round like hailstones or somewhat flattened like raindrops. This information had great potential to improve severe weather warnings.[18] Dual-polarization provides the NWS with three new and valuable products, namely differential reflectivity, specific differential type which allows for better prognosis of the impacts of light and intense rain, and the correlation coefficient which allows for the detection of wet or dry snow, large hail and even biological targets.

This technology has various other uses beyond NWS. For instance, one of NOAA's developments of radar technology is the mobile X-band dual-polarization Doppler radar (Hydro-Radar), which enables the study of storm dynamics, boundary layer turbulence and ocean-surface characteristics. A UC Santa Cruz scientist, working with meteorologists at the University of Oklahoma, is reported to use mobile storm-chasing radars, primarily developed for NWS, to follow swarms of bats as they emerge from their caves each night to forage for insects. The scientist uses data from 156 fixed NEXRAD weather radar stations around the country to produce images of bats appearing as distinct 'blooms' of radar reflectivity and give scientists clues to their behavior.

CONCLUSIONS

This chapter has dealt with federal public procurement in the United States outside defense and national security as it relates to innovation. The regulation binding on all federal agencies – the Federal Acquisition Regulation (FAR) – provides procedural rules that only control the procurement process. The two most important principles of FAR are cost savings and competitive processes. Priority for innovative products is not promoted as an essential policy consideration. Many federal agencies substitute FAR with similar rules better adapted to their own specific needs. Other important controlling documents are the executive orders issued by the Office of the President as binding instructions for federal agency officers and the OMB circulars and memorandums issued from time to time as instructions to federal agencies aiming at promoting economy, effectiveness and efficiency.

The substantive policies are mostly left to the discretion of the acquiring federal agency which is responsible for defining specifications and the evaluating criteria per procurement. That is to say, while federal public procurement is subject to a mesh of regulations issued and administered by a wide range of actors, significant decision-making power is left to the procurement manager in each agency. For larger, more diffuse agencies this empowerment is frequently transferred down to the specific procuring bureau. The independence of each bureau to come up with its own procurement needs and execution is based on the overall federal practice of giving maximum discretion to the end client or the acquiring agency in laying down procurement rules fitting to the overall mission.

Most of the 'strategic' procurement in the United States beyond defense and security is geared towards achieving social objectives such as environmental protection, energy conservation and assisting disadvantaged population groups. Innovation may be the outcome of the quest for such social purposes. Examination across a vast array of documentation for federal public procurement identified the closest link to innovation in terms of the development of environmentally friendly products and processes, the development of accessible technologies, and the improvement of space technologies. Cost-efficiency, a primary driver at least since public procurement reform in the early 1990s, also occasionally results in innovation. Beyond defense and national security, no regulation or established procedure in the United States provides guidance on PPI.

Procurement reform has become an important topic of public debate in the United States over the past couple of decades. Cost rationalization and minimization have surfaced as the most powerful drivers of procurement reorganization. The rationale of cost minimization notwithstanding, investing in innovation is a costly and risky enterprise that may occasionally run counter to short-term finance criteria. Discouragement of innovation may be reinforced by tightening financial management, which adds administrative burdens possibly prohibiting the most innovative (read: small) organizations from participating. Chief acquisition officers report tendencies to overspecify procured product characteristics mainly out of fear of litigation of unfairness claims. This runs counter to innovation where one would want to specify the desired performance characteristics and leave product characteristics to be proposed by the supplier.

Over the past few years a transformation has begun. The current administration has taken an interest in this area and has been pushing for changes. The new Office of Federal Procurement Policy (OFPP) of the White House has created committees with large and small government agencies to discuss innovative procurement practices. Infusions of managers from the private sector (and the defense contingent) to top public procurement management positions are argued to facilitate this process. There is significant thinking and awareness among federal public procurement managements regarding the need to keep improving the procurement process and promote innovation. However, innovation to them is rarely connected to the larger cause of improving the capabilities of the rest of the economy.

The issues of utmost importance for federal public procurement and its promotion of innovation include:

- Extensive litigation. Routinized legal challenge works against innovation by raising fears and undermining risk-taking behavior. A risky and uncertain process, innovation always involves some degree of adventure and cannot be specified with great detail in advance.
- Perils of overspecification. Detailed specification of product features may safeguard public officers from the perils of litigation, but works against innovation by limiting choice. In contrast, specification limited to customer needs, letting

suppliers decide how to satisfy these needs, works for innovation.
- Industry inclusion in decision-making. Involving industry early in the discussions over specific acquisitions (the FAA example) is expected to significantly improve the time, cost and innovative characteristics of procured products and services.
- Contracting. There is a lot of room for improving contractual agreements – in the direction of flexible contracts – in order to provide better incentives to suppliers to innovate, on the one hand, and improve the safeguards for public officers, on the other.
- People. There is a general understanding of lack of well-trained people in public procurement. As a countermeasure, it is supported that the inflow of people with industry backgrounds will improve both the efficiency and the effectiveness of the procurement process.
- Innovation awareness. There is little to indicate direct knowledge of procuring officers regarding the contribution of procurement budgets to the innovation capacity of the nation. The question should be asked of the specific acquiring officers. However, they do not have the bigger picture.
- Differences across federal agencies. Federal agencies will continue to be different in their procurement needs, processes and awareness of innovative solutions.

ACKNOWLEDGEMENTS

This chapter is based on the report 'Public procurement and innovation in the United States' carried out for Finpro, Project Advisory Services at the Finish Embassy in Washington, DC during the period 2010–2011. The study was carried out by Nicholas S. Vonortas, Pushmeet Bhatia and Deborah P. Mayer. Generous funding by Finpro is acknowledged. All remaining errors and misinterpretations are the responsibility of the authors.

NOTES

1. French public spending is expected to reach 57 percent of GDP, higher than any country in the eurozone (*The Economist*, 2013).

2. Estimates by Deltek (http://govwin.com/knowledge/government-contracting, accessed 24 November 2013). The total public procurement market is much larger, however, when one also considers the purchases of state governments. Overall, public procurement accounts for 10–15 percent of gross domestic product (GDP) in developed countries.
3. State governments are independent to create their own rules for the purchases of their agencies. For more detail see Vonortas et al. (2011).
4. The absence of regulation, such as the European Union (EU) legislation on public procurement (Directive 2004/18/EC and Directive 2004/17/EC), is apparent. No regulation or established procedure in the US provides guidance on PPI.
5. For instance, Sec. 6002 of the Resource Conservation and Recovery Act, which requires federal agencies to acquire as many recycled materials as practically possible. This Act has been encouraging both federal and state agencies in this direction.
6. The amazing resurgence of fossil fuels in the country during the past ten years or so is directly linked to the efforts by DoE to support new extraction techniques such as hydraulic fracturing (fracking).
7. For an excellent example of DoE's research and procurement strategies to promote innovative solutions with broad societal benefits, see the Offshore Wind Innovation and Demonstration initiative (OSWinD) (Vonortas et al., 2011).
8. See, for example, the independent acquisition information published by the Bureau of Ocean Energy Management and by the US Fish and Wildlife Service.
9. Several interviews with the chief acquisitions officers of four federal agencies were carried out for this study in order to identify innovative procurement practices in their agencies, and specific cases of procurement of innovative products and services from the private sector. The interviewed agencies included the Department of Energy, the International Trade Commission, the Department of Commerce, and the National Oceanic and Atmospheric Administration. Additional viewpoints were gathered during the workshop organized by the White House Office of Federal Procurement on 2 March 2011 that included federal agency procurement officers and industry representatives. For an analytical account of the interviews see Vonortas et al. (2011).
10. http://www.whitehouse.gov/omb/circulars_a130_a130trans4, accessed 5 May 2011.
11. The ITT Corporation's team includes AT&T, Thales, WSI, Corp., Science Applications International Corporation (SAIC), PriceWaterhouseCoopers, Aerospace Engineering, Sunhillo, Comsearch, Mission Critical Solution (MCS) of Tampa, Pragmatics, Washington Consulting Group, Aviation Communications and Surveillance Systems (ACSS), NCR Corporation and L-3 Avionics Systems and Sandia Aerospace.
12. http://oceanexplorer.noaa.gov/technology/subs/rov/rov.html, accessed 14 May 2011.
13. http://www.cio.noaa.gov/, accessed 14 May 2011.
14. NAO is North Atlantic Oscillation.
15. NAO 216–110 Eff: 5/19/06; Iss: 5/26/06.

16. NOAA. High Performance Computing Strategic Plan for FY2011–2015, 16 October 2008.
17. http://www.wdtb.noaa.gov/courses/dualpol/outreach/index.html, accessed 14 May 2011.
18. http://www.magazine.noaa.gov/stories/mag151.html, accessed 14 May 2011.

REFERENCES

The Economist (2013). Budgetary blues. September 28, pp. 48–49.

Edler, J. (2013). Review of policy measures to stimulate private demand for innovation. Concepts and effects. Report, Manchester Institute of Innovation Research, Manchester Business School, University of Manchester, and NESTA, January.

Edler, J. and Georghiou L. (2007). Public procurement and innovation – resurrecting the demand side. *Research Policy*, 36(7), 949–963.

Edquist, Ch., Hommen, L. and Tsipouri, L. (eds) (2000). *Public Technology Procurement and Innovation*. Boston, MA: Kluwer Academic Publishers.

Edquist, C. and Zabala-Iturriagagoitia, J.M. (2012). Public procurement for innovation as mission-oriented public policy. *Research Policy*, 41(10), 1757–1769.

Edquist, C. and Zabala-Iturriagagoitia, J.M. (forthcoming). Pre-commercial procurement: a demand or supply policy instrument in relation to innovation? *R&D Management*.

Fraunhofer (2005). Innovation and public procurement: review of issues at stake, final report. European Commission, DG Enterprise and Industry, ENTR/03/24.

Marx, R.W. (1984). Developing an integrated cartographic/geographic data base for the United States Bureau of the Census. *Proc. 11th International Cartographic Conference*.

OECD (2009). *Government at a Glance*. Paris: OECD.

Rothwell, R. (1984). Creating a regional innovation-oriented infrastructure: the role of public procurement. *Annals of Public and Cooperative Economics*, 55(2), 159–172.

Tomasi, S.G. (1990). Why the nation needs a TIGER System. *Cartography and Geographic Information Systems*, 17(1), 21–26.

Tomlinson, R. (1990). Geographic information systems: a new frontier. In Peuquet, D.J. and Marble D.F. (eds), *Introductory Readings in Geographic Information Systems*. London: Taylor & Francis.

Vonortas, N.S., Bhatia, P. and Mayer D.P. (2011). Public procurement and innovation in the United States. Report, Center for International Science and Technology Policy, George Washington University, Washington, DC.

7. Public procurement for innovation elements in the Chinese new energy vehicles program

Yanchao Li, Luke Georghiou and John Rigby

INTRODUCTION

China began to use public procurement as an explicit instrument of innovation policy (that is, public procurement for innovation or PPI) in 2006 when the National Medium- and Long-Term Program for Science and Technology Development (2006–2020) (hereafter MLP (2006–2020)) was announced. During 2006–2009 the central government launched further policy measures to implement PPI (for a detailed account see Li, 2011). The main (intended) approach to implementation was through 'innovation catalogues'. These were lists of innovative solutions, accredited by the Ministry of Science and Technology (MOST), contained within 'PPI catalogues'. They were to be produced by the Ministry of Finance (MOF) to guide government procurers in buying innovative solutions (Li and Georghiou, 2014). A second instrument, 'signalling catalogues', that is, lists of technologies that are identified as being much needed in China by the central government, were recognized as a complementary instrument to better link demand and supply (ibid.). This catalogue approach can be regarded as being what Edler and Georghiou (2007) termed 'general procurement', that is, an organized, 'routine' PPI mechanism where innovation becomes an explicit criterion in the tendering process.

The implementation of this approach in China has, however, come to a standstill, as a result of both domestic obstacles and international pressures. Domestically, a major challenge is created by the Chinese government procurement system, which is distinct from that of signatory countries to the World

Trade Organization Agreement on Government Procurement (WTO-GPA) in that it adopts a much narrower definition of government procurement (Wang, 2009; Li and Georghiou, 2014). The definition only considers procurement activities conducted by public organizations relying on fiscal funds (for example, procurement of office products and stationery by government agencies and public schools) as government procurement, while procurement activities conducted by state-owned enterprises and by some specialized ministries (for example, the former Ministry of Railways) are outside the scope of the Chinese Law on Government Procurement.

Our fieldwork has suggested that this regulatory arrangement significantly constrains the use of procurement to stimulate innovation, as many sectors that are likely to nurture innovations are beyond the authority of government procurers. Also, the fragmentation of the wider public procurement market in China (that is, public procurement which is within the scope of WTO-GPA but outside the scope of the Chinese government procurement system, taking a substantial proportion of China's public spending) allows serious regional protectionism to continue (Li and Georghiou, 2014). On the international side, China's drive towards 'indigenous innovation' (O'Brien, 2010) and especially the adoption of PPI, have led to deep concerns about China's protectionism from its major trade partners, in particular the USA (USCBC, 2010, 2011a, 2011b) and the European Union (EU) (EUCCC, 2011). After rounds of high-level discussions in July 2011, four key national policies underpinning innovation catalogues were abandoned (Li and Georghiou, 2014). The national-level PPI mechanism, which used the innovation catalogues approach, was officially abolished. Nevertheless, in the course of the fieldwork reported in this chapter, we identified other channels of PPI policies in China. These are summarized in Table 7.1.

One important channel is through national demonstration programs (see Table 7.1, last column), that is, lead market initiative (LMI) type measures (CSES and Oxford Research, 2011), which is a mix of demand-side innovation policy instruments to 'pull' and accelerate the commercialization and market transformation processes. The procurement elements of these programs could be within or outside the scope of China's narrowly defined government procurement regulations, depending on whether government agencies are the end-users or not. Procurements stimulated by these programs can be regarded as being what

Table 7.1 The range of national policies promulgated for innovation procurement

Forms	Routinized mechanism via accrediting catalogues	Signalling catalogues of equipment and other strategic technologies	Demonstration programs for strategic and emerging areas
Rationale	Enhancing communication between suppliers and procurers	Signalling national demand; technology roadmapping	Creating lead markets; systemic mix of policy instruments
Implementation (based on fieldwork)	Ambiguous national measures; regional autonomy in developing local mechanisms; diversified across regions	Governmental departments regularly launch catalogues to inform suppliers; strategic procurement takes place from time to time	Targeted at various sectors, for example, LED lighting, solar energy and new energy vehicles (NEV)
Current status	Withdrawn in July 2011 in response to international concerns	Relatively on track; facilitated technology advancement in a variety of ways	Results diversified across sectors; achievements as well as challenges

Source: Li and Georghiou (2014).

Edler and Georghiou (2007) termed 'strategic procurement', which 'occurs when the demand for certain technologies, products or services is encouraged in order to stimulate the market' (p. 953). These programs are targeted at promising and strategically important sectors, for example, the sector we look into in this chapter, new energy vehicles (NEVs).

This chapter undertakes an exploration of policies for procurement of innovation, with the main aim being to bring to light the details and difficulties involved in implementing the policy in China. In particular, an examination is made through the lens of city-level procurement activities stimulated by central-level innovation policy initiatives in China. The particular case examined is that of NEVs, for which we gathered both primary data (interviews) and secondary data (policy documentation). We firstly outline the context of the Chinese NEV program, and then move on to investigate cases of NEV procurement in two participant cities. We conclude by discussing some of the key issues identified from the cases.

THE NEV DEMONSTRATION PROGRAM IN CHINA

According to the scope defined by the Chinese government,[1] NEVs include hybrid vehicles, plug-in hybrid electric vehicles, battery electric vehicles and fuel cell vehicles. Research and development (R&D) on NEVs in China has been supervised by the MOST since the early 1990s, mainly in the form of high-tech R&D programs (Sun, 2012). In China, the development of NEVs has been considered necessary for several reasons. As a result of the high growth rate of the economy, the demand for vehicles has been increasing quickly. The country suffers from a severe energy shortage and environmental pressure (World Bank, 2011). China has made a commitment to the United Nations to reduce its carbon emissions by 40–45 percent, the development of NEVs being recognized as an important way of realizing this target (State Council, 2011a). Meanwhile, although China is now the largest and fastest-developing market for vehicles in the world in terms of both manufacturing and sales (Lin and Wang, 2012), engine-related technologies have been imported from developed countries and controlled by multinational automobile suppliers, while domestic firms occupy only a small share in the traditional vehicle market (ibid.). Therefore, the government is determined to capture the opportunity of developing new types of vehicles and further escalate the automobile industry (State Council, 2012).

After years of R&D support, the central government more recently considered it timely to facilitate the commercialization of NEV technology. Major suppliers had developed their prototypes, which were in need of market access (Gong et al., 2013). Since 2009, a variety of innovation policies have been announced to support NEV commercialization, covering not only the supply side but also the demand side of the market. This engaged additional government agencies (beyond the MOST) along the supply chain. In Table 7.2 we briefly classify the key NEV-related innovation policies adopted in China to present our overall

Table 7.2 Classification of NEV-related innovation policies in China

Overall policies		Supply-side policies	
Policy type	Authority	Policy type	Authority
Implementation measures for indigenous innovation strategy	State Council	R&D programs for example, 863 projects (the National High-tech R&D Projects)	MOST
Stimulating policies for emerging technology sectors	State Council	R&D funding for public institutes, universities and state owned enterprises	MOST
Development plan for energy saving and new energy vehicles (2012–2020)	State Council	Networking measures for example, alliances, incubators and training	MOST, MIIT

Demand-side policies		
Category	Policy type	Authority in charge
NEV program focused polices	Overall measures detailing implementation approaches	MOF, MOST, MIIT, NDRC
	Funding/subsidies related measures	MOF
	Catalogues of approved NEV models	MIIT
	Regulations on other issues for example, safety, infrastructure building	MOF, MOST, MIIT, NDRC
Regulations	Regulations on tax reduction, NEV supplier qualification, emission, and government procurement procedures for NEV	National Bureau of Taxation, MIIT and MOF
Standards	Standards on oil-saving rate testing and charging facilities etc.	MIIT

Sources: Derived from policy analysis. Policies were found on websites of the State Council, related ministries, and the policy section on the China NEV website http://www.chinaev.org/DisplayView/Vip/Policy/Index.aspx (accessed February 19, 2013).

understanding of the policy context. The corresponding institutional set-up and division of labor were elaborated by Liu and Kokko (2013). The most systemic policy measure is the Energy-saving and New Energy Vehicles Demonstration, Promotion and Application Program[2] (hereafter 'the NEV program'), which aims to create lead markets for NEVs in selected cities.

Initiated jointly by the MOF, the MOST, the Ministry of Industry and Information Technology (MIIT) and the National Development and Reform Commission (NDRC) in January 2009, the NEV program aimed to promote the use of around 1000 NEVs in each of a series of selected cities during 2009–2012 (MOF et al., 2010b). In total 25 cities were selected as participants for the public sector demonstration, whereby government agencies or public transport companies (which are state-owned) were to be given subsidies when purchasing buses, taxis, government cars, environment maintenance vans and mail delivery vans using NEV technology.

Both cities investigated in this chapter, Jinan and Shenzhen, were participants throughout the period 2009–2012. In addition, Shenzhen was one of the six participants selected for the consumer NEV demonstration (MOF et al., 2010a), which allowed citizens in Shenzhen to enjoy subsidies when purchasing NEVs for their own use.

According to a ministerial-level interviewee who was involved in the program, the selection of participant cities was conducted on the basis of proposals submitted by the candidate cities. Cities were required to design realistic implementation plans and ensure that they fulfilled the goals by the end of 2012. Expert groups were then organized by the ministries to carry out fieldwork to evaluate the potential of candidates. Selection criteria included the size of the local market, the status of the local automobile industry and financial conditions of local governments as well as a consideration of the national industry strategy. Geographically, most of the selected cities were located in the eastern and middle part of China, and were mainly municipalities, provincial capitals or medium-sized cities. For some cities that had relatively small local markets but good industrial potential, the option exists to form 'city clusters' to participate (Gong et al., 2013). For cities enjoying the private consumer subsidies from the MOF, additional criteria needed to be fulfilled, such as consumption capacity of citizens and traffic conditions in the locality. The amount of the subsidies was set according to the overall price difference between NEVs and traditional vehicles with similar performance characteristics (MOF and MOST, 2009). Both provincial and city-level governments were obligated to provide additional subsidies to procurers and to provide special funding for infrastructure construction and maintenance (ibid.).

The ministries also stipulated that, to be subsidized, procurers had to choose NEVs from the *Catalogues of Recommended Vehicle Models for NEV Demonstration Program* produced by the MIIT (MOF and MOST, 2009; MOF et al., 2011). Detailed criteria for products included: the oil-saving rate of hybrid cars had to be above 5 percent compared to traditional vehicles with similar performance characteristics, while the oil-saving rate of hybrid buses had to be above 10 percent; the warranty of batteries and other key segments provided by manufacturers had to cover three years (or 150 000 kilometres) or longer; the manufacturing capacity of key components suppliers needed to reach a certain threshold (MOF and MOST, 2009). Meanwhile, the procurers were required to organize a public tendering process to buy NEVs with clear specifications of the model, quantity, price and after-sales services (ibid.). We illustrate the design of the NEV demonstration program in a simplified way in Figure 7.1.

Following the central-level initiative, participant cities have had the flexibility to design their own form of implementation. They demonstrated strong enthusiasm to promote NEVs, since each city had its carbon emission reduction task allocated from the central and provincial governments (State Council, 2011b). However, cities with or without advantages in terms of traditional automotive technologies also wanted to take the opportunity to develop local NEV industries as an instrument of economic development (Gong et al., 2013). Although the original target set by the central government was 1000 NEVs for public use per city, most of the participants set up their own targets at a level far higher than this. For example, Beijing (with a population of approximately 20 million) aimed to promote the application of 5000 NEVs by the end of 2015, and Shenzhen (with a population of approximately 10 million) aimed to promote 24000 NEVs by the end of 2012 (CATARC, 2010). A wide range of policy instruments was used by localities. Besides typical measures such as consumer subsidies, tax reduction and public procurement, some cities issued favorable policies for NEV users to reduce their parking, toll and electricity fees; or in the case of

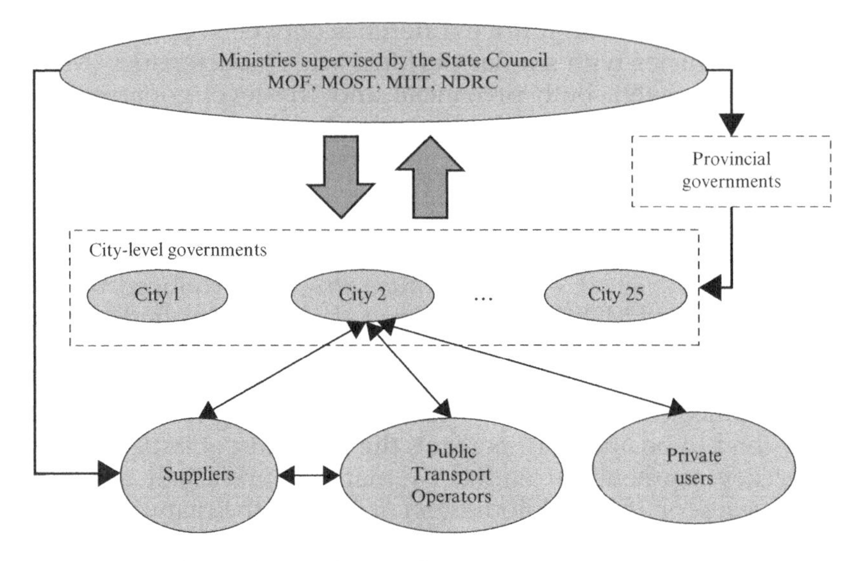

Sources: Based on policy analysis and interviews.

Figure 7.1 The design of the NEV program

Beijing, where people had to take part in a lottery to get a license plate for their new cars, NEV buyers did not have to go through the competition process. Some provinces followed the systemic approach adopted by the central government and initiated their own demonstration programs in their region to encourage more cities to become involved rather than just those selected by the ministries (Gong et al., 2013).

The progress of participant cities has been uneven, and none have in fact achieved their three-year targets (Gong et al., 2013) at the time of writing in October 2013. On average, only 38 percent of the targets set by cities for public use were realized by 2011 (Huang et al., 2012). The overall fulfillment ratio for both public and private uses was as low as 26 percent, primarily because of the unrealistic goals set by the cities in the first place (Gong et al., 2013). The actual quantity of NEVs promoted through the program during 2009–2012 was 27 400,[3] approximately 23 000 of which were procured by public bodies and only 4400 NEVs purchased by private consumers. In terms of production, there were more than 400 NEV models produced by 76 manufacturers listed in the *Catalogues of Recommended*

Vehicle Models for NEV Demonstration Program by October 2011 (Gong et al., 2013), although many of the suppliers were not yet capable of large-scale manufacturing (ibid.). The number of NEV charging and battery swapping stations and charging stations in China was 174 and 8107, respectively, by the end of 2012 (see the previous note). Although many issues such as regional protectionism were manifested during the demonstration program (as illustrated in Gong et al., 2013, and as we shall see in the case studies) and impacts in the longer run remain to be seen, our interviewees considered the program fairly effective in terms of raising stakeholders' awareness and promoting technological advancement.

METHODOLOGY TO BUILD THE CASES

The cases presented in this chapter form an integrated part of a broader, exploratory study (Li, 2013) on PPI policies and practices in China. They were built on the basis of a comprehensive analysis of the Chinese context in terms of the innovation system, the procurement system, and the identified PPI policy channels (as shown in Table 7.1). We distinguish three levels of governance: the macro, national level of policy-making; the meso, regional level of policy articulation; and the micro level of policy implementation. In particular, understandings of the policy processes at upper levels have been obtained primarily through documentation analysis supplemented with interviews with national and regional officials from science and technology (S&T), industrial and financial departments, while understandings of the implementation processes at the micro level (that is, the level where the two cases locate) have been obtained primarily through semi-structured interviews with various practitioners and stakeholders including public procurers, local government officials, suppliers and (occasionally) end-users.

It should be noted that across the participant cities there have been many procurements stimulated by the NEV program.[4] We illustrate these by analysing two cases that took place in the cities of Jinan and Shenzhen. Primary data collection was carried out between December 2010 and May 2011 in both of the cities as well as Guangzhou (the capital of Guangdong province where Shenzhen is located) and Beijing (to gain insights from ministerial officials regarding the two cities' practices). To build the cases ten interviews were conducted, of which two were with national officials in charge of the NEV program, two with local officials from the city governments (one of them was meanwhile a user of electric sedans), four with different NEV suppliers, and two with public transport operating companies (that is, procurers and users of coaches). For national officials, questions regarding the broader picture (for example, the two cities' overall progress, competitive advantages and weaknesses) were asked; for micro-level practitioners, questions regarding the procurement cycles (that is, demand articulation, selection procedures and criteria, stakeholder interactions, contract delivery, outcomes and lessons) were asked. We structure the cases into three parts, that is, the stage before the procurement, which we title as the 'pre-procurement' stage, the procurement process, and a short discussion about the outcomes and impacts.

CASES OF PROCUREMENT

Jinan: Procurement of Hybrid Coaches for National Games 2009

Jinan[5] is the capital of Shandong province situated on the eastern coast of China. Historically it has been an important city in China in terms of its economy, culture and transport. In 2011 Jinan ranked twenty-third in Chinese cities' gross domestic product (GDP) ranking.[6] According to the materials Jinan submitted to the ministries, its local administrative set-up for NEV demonstration is a group led by the mayor with heads of the local Development and Reform Commission (DRC), Bureau of S&T, Bureau of Finance, and Commission of Economy and Informationalization as group members. These organizations also set up a specialized office to monitor implementation. The main policies in the locality related to the NEV program include Shandong Provincial Government (2009a, 2009b, 2009c, 2009d) and Jinan City Government (2009a, 2009b), and the Implementation Measures on NEV Demonstration and Promotion in Jinan approved by the four supervisory ministries.[7] Content analysis suggests that these policies are in general articulated based on the approach adopted by the central government (see Table 7.2), taking local circumstances (for example, challenges and opportunities faced by the local NEV industry) into account.

Some electric vehicle (EV) manufacturers in Jinan and other parts of Shandong have been focused on producing low-speed vehicles for rural areas, and their products are very popular among farmers. However, thus far the ministries have not announced any supportive measures for low-speed EVs, as many experts do not consider low-speed EV technology based on lead-acid batteries as a promising technology, since it can be very harmful for the environment.[8] In this respect Jinan is trying to shift its industrial focus from low-speed EVs to leading-edge NEV technologies. The city government is trying to attract NEV key components suppliers to invest in the locality by offering access to a market created by the demonstration project (interview R4O_NEV). The NEV industry in Jinan has been in the nurturing stage since 2005 when the first battery factory was founded, and now several domestic automobile suppliers have opened local branches there.[9]

The pre-procurement stage: the need

As with the Beijing Olympics 2008, Shanghai EXPO 2010 and Shenzhen Universiade 2011 (see the section on 'Shenzhen: Procurement of NEVs for Universiade 2011'), the National Games of China in 2009 provided a good opportunity for the host city, Jinan, to improve its public transport infrastructure and demonstrate the use of NEVs. The Jinan government normally allocates 60 million yuan per year to the local public transport company (state-owned) as operation subsidies. In 2009 it decided to provide additional funding of around 40 million yuan to conduct public procurement of a batch of NEVs, and hence to support their use during the National Games and to kick off the implementation of the NEV demonstration program (interviewee R4O_NEV). The Jinan government set up technological requirements jointly with the public transport company. They required that the coaches should be 12-meter-long diesel–electric hybrid models with paralleled batteries, and their exhaust emissions should be less than China's national Tier IV standard.

The procurement process

With a total budget of 100 million yuan, coupled with subsidies from the central government, the Jinan government procured 100 hybrid buses on behalf of the operating company for the National Games of 2009. These NEVs operated between sport venues, coach stations and athlete hotels during the game, and

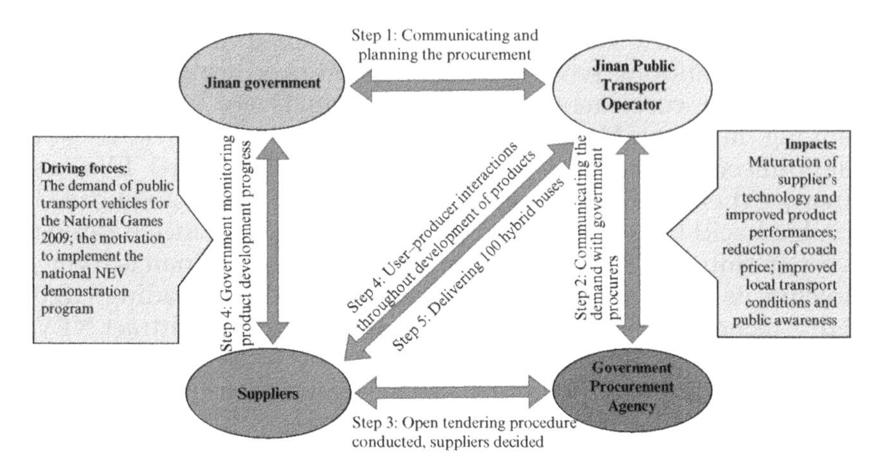

Figure 7.2 The procurement process of NEVs for the National Games 2009

served as regular buses on four of the public transport lines in Jinan afterwards. The overall process of procurement is illustrated in Figure 7.2.

According to the interviewees in Jinan, the government decided to buy hybrid coaches rather than electric ones for three reasons. The first was that hybrid vehicle technologies were more mature than EVs in early 2009 when the demonstration program had just begun. The second reason was the budget issue: each pure electric coach cost around 1.2 million yuan at that time (1.8 million yuan minus the national subsidy of 0.6 million yuan), while each hybrid coach cost around 0.95 million yuan (1.2 million yuan, minus the national subsidy of 0.35 million yuan). A third reason was that the locality was not able to build charging infrastructure for EVs in a short time with a limited budget (one charging station cost around 30 million yuan), and hence hybrid coaches were a good option.

The operating company then published an invitation to open tendering via the Shandong Government Procurement center, and nine manufacturers (qualified by the MIIT) submitted their bids. R4O_NEV (a vice-director of Jinan S&T Bureau) said the requirements they set were considered too high by most of the bidders at that time, and only two companies eventually provided acceptable product designs. Company A, a firm from Shandong province, obtained the top score and won a contract for 80 coaches with a value of 87.4 million yuan; and Company B, from outside the province, achieved the second-highest score and won a contract for 20 coaches. Both companies signed the contracts in March 2009, and the deadline for delivering the coaches was the end of July 2009 as the National Games was starting in October.

Interviewees in both the Jinan government and the public transport company admitted that they preferred to buy products from company A, since the provincial government would provide its own subsidies (in addition to national subsidies) for the purchasing of NEVs from local companies (that is, situated in the province). If possible, they would even prefer to purchase from a supplier situated in the city, in the hope that procurement activities may contribute to the development of a local NEV industry.

Company A asked for an extension of the deadline to the end of September 2009, but despite this, it failed to manufacture all the needed products before the Games. Under the pressure of delivering products on time, it substituted the original key components (including the engine, the controller and the battery) with imported, good-quality alternatives to meet the requirements of the contract. According to national and provincial support policies for the demonstration program, only domestic products with 'indigenous IPRs' (intellectual property rights) could enjoy the subsidies, that is, in this case at least two of the three key components should be designed by native companies. Therefore, in order to fulfill the contract requirement and to get the subsidies, Company A spent the following months seeking to improve its products and gradually substituted the imported components with its own, improved products (key IPRs fully owned by company A). By early 2010, all 80 coaches had been equipped with domestically made components.

The procurer interviewee indicated that the prototype provided by Company A was qualified to enter the market according to the MIIT criteria. However, the manufacturing capacity of Company A was limited at the time, making the delivery of products on time difficult; and indeed the operating company was concerned about using such new products immediately, during the period of the Games. Jinan was one of the first cities procuring NEVs in the country after the start of the NEV program, and the national government was also concerned about the performance of the NEV technology. One interviewee

(R4P_NEV, a manager in the Jinan Public Transport Group) noted the inexperience of both parties to the contract: 'we were not experienced as procurers and they were not experienced as a supplier'. In consequence, the three parties agreed to extend the deadline of product delivery, and to temporarily replace the core components with imported ones to lower the risk and guarantee product performance.

By the time of interview (May 2011), the 80 coaches had been operating in the public transport lines of Jinan for more than one year. A manager in the operating company (interviewee R4P_NEV) believed that the quality of these coaches was very good, with an overall oil-saving rate of 26 percent (while the threshold for subsidizing was 20 percent). The performance of the key components was stable and the hybrid coaches could now operate as frequently as traditional vehicles at an overall rate of 98 percent.

Although the procurer company adopted an open tendering procedure for the procurement, it is worth noting that the operating company and Company A (both are state-controlled companies in Shandong province) had been in a co-operative relationship for a large number of years prior to the NEV program. According to interviewee R4P_NEV, products of Company A account for one-quarter of the total number of coaches in the Jinan Public Transport Group. One major reason for this co-operation is that Company A is located close to Jinan, and hence it can provide after-sales services more easily. A manager of the province's procurement team (interviewee R4F) said that he preferred company A as the supplier, because it was very familiar with the traffic conditions in Jinan and could follow the national and industrial changes quickly to satisfy the changing needs of customers. This long-term relationship provided a basis of trust for the procurer to choose Company A.

At the end of 2010, the operating company published another tendering invitation for 100 hybrid coaches. Company A won the contract, again due to its previous experience. This time it submitted the bid at a lower price (around 900 000 yuan per coach), while other companies failed to provide competitive offers.

A manager in the operating company mentioned that it had raised the testing standards for the second batch of coaches by ordering a prototype and conducting a comprehensive examination with technological experts. They made suggestions regarding battery configuration (the number and series of batteries were changed according to predicted routine traffic conditions) and a higher oil-saving rate was realized. Interactions between the user and the producer helped to first identify problems and then find a solution that better satisfied the customer demand before the model was put into larger-scale production.

Outcomes, impacts and issues

One impact of this procurement was the maturation of Company A's technology and the improvement of product performance. As mentioned, Company A initially failed to deliver satisfactory products on time for the National Games 2009, and imported key components were used instead. The company then improved its products and replaced the imported components with its own by early 2010. In the second procurement, user–producer interactions facilitated further improvement of Company A's products. Another impact was the reduction of the coach price. In the second procurement, Company A submitted an offer that was around 25 percent lower in price than the first batch, beating other bidders in the tendering. A third impact was that the two procurements improved the conditions of public transport in Jinan to a certain extent, and improved public awareness of NEVs in the locality as well, since the four transport lines which the hybrid coaches are used on cover very popular routes in the city center.

The procurements in this case did not by themselves have much impact on the building of Jinan's local NEV industry, as the scale was rather small. It did facilitate incrementally gaining access to a wider market for the supplier. A recent search of secondary data in May 2013 suggested that, following the delivery of the second contract in this case, Company A won more contracts from the Jinan government as well as other cities in Shandong province; beyond Shandong, Company A has been gradually overcoming barriers created by regional protectionism and winning small-scale contracts (normally under 100 NEVs). A sales milestone of 1500 NEVs was achieved in early 2013, which is already a number coveted by most of the Chinese NEV suppliers.

However, despite Jinan being one of the first-batch participant cities and having conducted one of the earliest procurements of NEVs, it is now the slowest and the least motivated among the first-batch participants (Huang et al., 2012). Delay could result from a wish to avoid risky commitments in the face of policy uncertainty. The motivation to carry that risk is reduced as Jinan lacks local industrial support, and hence has a reduced economic incentive to bear it.

Shenzhen: Procurement of NEVs for Universiade 2011

Our second case is located in Shenzhen. Situated in the Pearl River Delta, Shenzhen is the second-largest city (after the provincial capital, Guangzhou) in Guangdong, and the first special economic zone in China nominated by the State Council in 1980. It has developed from a small town into an internationalized city during the past three decades, and it is well-known for its impressive 'Shenzhen Speed',[10] a label that signifies the technologically advanced character of the people and the area. Shenzhen is ranked fourth in the 2011 GDP ranking (after Shanghai, Beijing and Guangzhou) of cities in mainland China, and ranked as the most innovative city in China by Forbes.[11] The governance of the city is similar to that of Jinan, that is, an administrative group led by the vice-mayor with heads of the local Development and Reform Commission (DRC), Bureau of S&T, Bureau of Finance and Commission of Economy and Informationalization. Main policies in the locality related to the NEV program include Guangdong Provincial Government (2009, 2010), and Shenzhen City Government (2009a, 2009b, 2011).

BYD[12] and Wuzhoulong[13] are the two major NEV manufacturers in Shenzhen. Founded in 1995, BYD started its business in battery manufacturing. In 2003 it purchased Xi'an Qinchuan Automobile factory and entered the car manufacturing and sales business. In 2008 it purchased the Ningbo Zhongwei Semiconductor factory and integrated the upper-stream supply chain of motor manufacturing for electric cars. In July 2009 it also purchased Changsha Meidisanxiang Coach Co. Ltd and through this acquired acceptance from the MIIT for manufacturing coaches. Now it is a privately owned high-tech enterprise with information technology (IT) and automobile businesses. The main advantage of BYD's NEV business compared to its competitors lies in the fact that almost all the upper-stream suppliers are from the BYD group, whereby they can control product price and maintain a maximum profit (interviewee R3S_NEV, a manager from Wuzhoulong Co. Ltd). BYD has adopted a localization strategy in many regions. In 2011 it had nine R&D and manufacturing bases across the country, including Guangdong,

Beijing, Shaanxi and Shanghai. BYD's headquarters are now in Pingshan District in Shenzhen.

In contrast to BYD, which focuses on battery and electric vehicles, Wuzhoulong Motors has had a clear strategy of developing energy-saving coaches since it was founded in 2000. Its core advantage lies in its vehicle material manufacturing technologies and its smart hybrid motor controller, which integrates an energy controlling system, automatic clutch controlling system and information management system. Approved by the Shenzhen government in 2005, seven hybrid coaches manufactured by Wuzhoulong became the first demonstration public transport line in the country. The average oil-saving rate of Wuzhoulong's products was 25–30 percent.

The pre-procurement stage: a pilot project

A pilot project was initiated by the Shenzhen government in 2008 to demonstrate the use of hybrid buses and electric cars in typical urban traffic conditions. This project was then approved by MOST as part of the 11th Five-Year 863 Key Program on Energy Saving and New Energy Vehicles, in order to provide a reference for further country-wide NEV demonstration. The budget provided by the Shenzhen government was 50 million yuan to cover the expense of purchasing and maintaining 30 hybrid buses and 20 electric cars. Wuzhoulong and BYD were chosen as the suppliers for the hybrid buses and electric cars, respectively. Three public transport lines were designed with ten hybrid buses in each of them. Twenty electric cars were then allocated by the local DRC to governmental bodies (for example, the S&T Bureau, Environment Protection Bureau and Transport Bureau), with charging and testing facilities installed accordingly.

According to interviewees, the procurement of hybrid buses was straightforward, while for electric cars the project group decided to adopt the mode of 'government renting' instead of 'government procuring', as the government bodies at that time had failed to get permission for the procurement from the local Bureau of Finance. The car model rented was an early version of F3DM,[14] a plug-in hybrid compact sedan. The rental fee was fixed at 80 000 yuan for two years, which, according to interviewee R3S_NEV, was much lower than the real cost, as the supplier provided a whole package of post-rental services, insurance and tax, and so on. Still, the supplier was very actively involved and grateful for this opportunity to promote its prototype.

User–supplier interactions were frequent during the two years. BYD set up a specialized service group for the project which had two major functions: a routine visit to the user organizations to collect feedback and a trouble-shooting service when they encountered any problems. Users were requested to provide feedback regarding vehicle functionalities. Although radical technological change was not stimulated by those interactions, many detailed improvements regarding product performance were realized. One example was that the dashboard, which used to be similar to traditional ones, was redesigned into a more user-friendly version displaying the dynamic battery level and operating mode, and so on. A 'low-carbon' version of F3DM was developed after this project, integrating a solar energy panel on top of the car and hence further saving energy.

The impact of this pilot project lies in the growing interest and familiarity among citizens in Shenzhen with NEVs, as government bodies volunteered to use them. The number of telephone enquiries about F3DM received by BYD increased significantly due to the demonstration effect. Early users were open enough to introduce their experience to the public. One of them (interviewee R3U_NEV) indicated that he was impressed with two things in particular: one was that the acceleration speed was very high, so the driving experience was superior; the other was that by using the new car he paid 80 percent less for energy costs, including electricity and petrol.

F3DM became well known, gradually gaining ground in the private market afterwards. By May 2011, BYD had sold around 600 units of this model to private consumers, the majority of whom came from Shenzhen city. Since Shenzhen was one of the demonstration cities for private user subsidies, the price of F3DM for local consumers was 80 000 yuan after the subsidies from central and local governments (the subsidizing amounts were 50 000 yuan and 30 000 yuan, respectively, covering 50 percent of the total price), compared with 169 800 yuan for customers from other regions. The project provided an opportunity for the government, the suppliers and the users to interact with each other, and hence build a relationship which led to the procurement of NEVs for the Universiade 2011 in Shenzhen.

The procurement process

The Universiade hosted by Shenzhen city in 2011 provided a major opportunity for the locality to implement an NEV

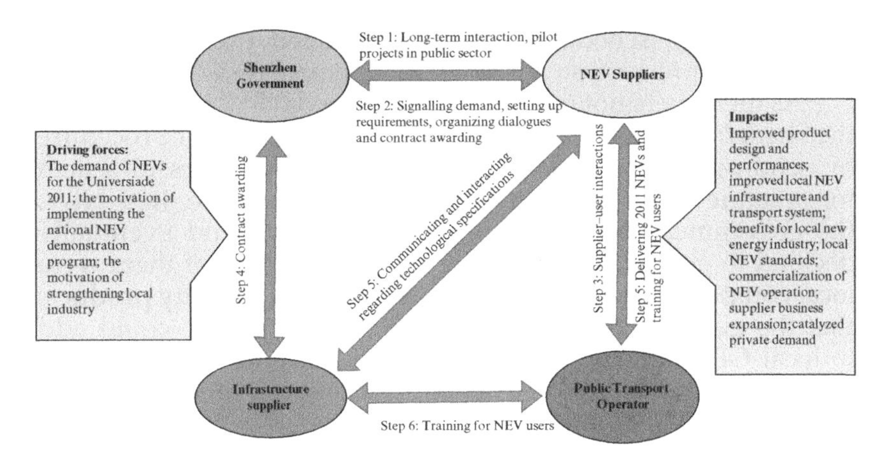

Figure 7.3 The procurement process

demonstration program through constructing the infrastructure and providing public transport. Since Shenzhen is well known as an innovative and active city, the local government set 'green technology' as one of the themes, and the NEV policy was one way of demonstrating this. This fortunately coincided with the national NEV demonstration program. The number of NEVs in Shenzhen increased significantly up to 2011 as a result of public procurement, exceeding the combined number in the Beijing Olympics in 2008 (around 500 NEVs) and Shanghai EXPO in 2010 (around 1300 NEVs). The procured vehicles covered a range of NEV types including hybrid, fuel cell and battery-supplied. The overall process of procurement is illustrated in Figure 7.3.

Performance requirements (including oil saving rate, charging speed, driving range and maximum speed) for the NEV in Universiade were set by the Shenzhen government in their announcement of the Implementation Plan for NEV Demonstrating Operation During the 26th Summer Universiade. The top priorities were indicated to be 'safety, environment protection and a demonstration of the Shenzhen characteristics'. Four aims were outlined in the document: to get NEVs to make up more than half of the public transport; to cover the sports venues using a structured traffic network; to realize diversification of various NEV technologies; and to explore a commercialized way of operating NEVs. The government also specified technological configuration requirements with a clear manufacturing schedule, and an organized expert group to monitor the production and construction progress.

As there were not many commercialized NEVs in the market, the Shenzhen government adopted a restricted tender procedure by sending out invitations to well-known suppliers to search for qualified products. According to interviewee R3S_NEV, the government compared available products and eventually decided to go for BYD and Wuzhoulong, as both of them were local companies and capable of providing good-quality products that the government knew about, and their income could add to local GDP growth. It was a great opportunity for Shenzhen to demonstrate its innovativeness by using a local product that was nationally leading-edge. The relationship between the government and the two local firms was built soundly from the early pilot project stage and other forms of communications; for example, interviewee R3S_NEV2 mentioned that whenever they developed new models, they would inform the government in order to see whether there were opportunities to use them.

Available sedan models provided by BYD included the plug-in hybrid car F3DM as mentioned above, and a newly developed pure electric model E6,[15] both of which have been listed in national and local innovation catalogues, and in the catalogue of recommended NEV models produced by the MIIT. By May 2011, a batch of 50 E6 cars had been operating in Shenzhen as taxis for a whole year with a total mileage of 3 million kilometers. The electric coach model produced by BYD was K9, which was pure electric with a fast charging function (50 percent of the capacity can be charged within 30 minutes), and a solar energy panel on top to provide additional electricity. Based on BYD's testing result, K9 (with air conditioners turned on) consumed less than one-third of the energy used by a traditional coach.

Wuzhoulong proposed three models of NEVs. One was FDGFCL10, a ten-seater hydrogen fuel cell van with a maximum speed of 40 km/h, featuring a smart inductive electricity assisted steering system, 'stepless' driving system and a permanent magnet synchronous motor. A second model was FDG6120SDEG, a hybrid coach model equipped with lithium iron phosphate battery featuring an automatic series-parallel hybrid driving system, stepless series-parallel transmission function, automatic mechanical transmission technology and a diesel engine that met the requirement of the Euro-III emission standard. The ratio of electric power could reach 43 percent and the oil-saving rate was above 30 percent. The third model was FDG6700EV, a pure electric coach model equipped with an engine produced by Shanghai Dajun, and a lithium iron phosphate battery.

Of the 2011 NEVs procured for the Universiade, BYD provided 200 pure electric buses (K9) and 300 pure electric taxis (E6), and Wuzhoulong provided the rest, a total of 1511 energy-saving and new energy buses. Based on the three technological models described above, Wuzhoulong designed six different types of vehicles for Universiade use, including 1350 hybrid single-layer buses, 20 hybrid double-layer buses, 53 pure electric buses, 26 pure electric vans, 60 hydrogen fuel cell sports venue vans and two hydrogen fuel cell coaches. Of these vehicles, the double-layered hybrid coaches and pure electric vans with exchangeable batteries were new to the country. All these NEVs were allocated to 77 public transport lines specialized for Universiade, covering all the sports venues and constituting an NEV transport network in Shenzhen.

The other part of the procurement was infrastructure construction. Shenzhen Bureau of Power Supply commissioned a local firm, Shenzhen Lineng Charging Station Co. Ltd (part of the Putian Group, which is controlled by the State-Owned Assets Supervision and Administration Commission of the State Council, SASAC), to conduct the work, and built the largest charging station thus far in the country, the Shenzhen Universiade Centre EV Charging station. The work included two parts: upgrading work for 25 existing charging stations that were already in use in Shenzhen before the Universiade; and building 34 new charging stations including 22 fixed stations with 161 fast-charging ports and 12 mobile stations with 125 charging ports. In total, the 59 charging stations were sufficient to provide power for all the NEVs during the event. Interviewees were unwilling to disclose what the Shenzhen government had spent on these vehicles and infrastructure.

Outcomes, impacts and issues

The first important outcome of the procurement was improvement of product design and performance. In response to the requirement set by the government, Wuzhoulong designed a two-layered hybrid bus model that was novel in the domestic market. The LED display system in the bus was improved due to interactions between users (drivers) and the supplier. The system now can display the vehicle's dynamic operating status according to the user's set-up options and traffic conditions. There are different levels of authority for set-up options, for example regular options for everyday drivers and advanced options for engineers.

Another important outcome was that the local NEV infrastructure and public transport system were improved, mainly as a result of the event. The 2011 NEVs became part of the main transport system in Shenzhen. According to Shenzhen's plan in 2009, the city aimed to promote the use of 24000 NEVs (both public and private sectors) in the locality by the end of 2012, and build 50 charging stations for buses, 2500 charging piles for government vehicles, 200 charging stations and 30000 charging piles for the public. This procurement for Universidade provided practical experience for later-stage construction, and it considerably accelerated Shenzhen's progress towards achieving the target for the NEV demonstration program.

There have been benefits for the local new energy sector as a whole. Shenzhen is now building a national new energy industry base to promote nuclear, solar, wind and bio-energy technologies and electricity storage technology. The development of NEVs has provided an opportunity for the supply chain to explore the possibility of interdisciplinary innovation.

This procurement meanwhile contributed to NEV standardization in the locality. In 2010 the government organized infrastructure suppliers, including China Southern Power Grid Co. Ltd, to conduct research on existing standards and developed their Technological Standards of EV Charging System in Shenzhen and Standards of Monitoring System on NEVs in Shenzhen, making Shenzhen a pioneer in the country exploring a unified system of standards.

The demonstration during Universiade also attracted capital from the public to invest in NEV promotion. BYD, Wuzhoulong, Shenzhen Public Transport Group and China Putian Group have signed a contract together to promote NEVs in a more commercialized way by applying for commercial loans. For future construction work, the government only provides funding for hybrid bus charging stations, while charging stations for taxis, government cars and private cars will be built using capital raised from public sources.

The diffusion of products by BYD and Wuzhoulong has been effectively enhanced. Suppliers in this case received many invitations to tender from other regions and countries, since their products performed very well for the event and their prices were competitive. Suppliers' collaboration with domestic and international business partners was enhanced as well.

There are other impacts that we have noted in the case: for example, product costs were lowered effectively, and public awareness of Shenzhen's NEV products was enhanced. Shenzhen gathered valuable experience for the NEV commercialized operation, and it was considered by the ministries as the pioneer among participant cities, winning the prize of the Annual (2011) Best Participant City in the NEV Demonstration Program.[16]

DISCUSSION

This chapter has examined two cases regarding city-level public procurement activities stimulated by the central-level NEV program initiatives. The cases were chosen to illustrate the characteristics and issues associated with concrete PPI processes. Table 7.3 summarizes the two cases, indicating both similarities and differences. Commonalities were the incremental nature of the technological development, and effective import substitution. Nonetheless, this was innovation procurement in the sense that the suppliers were set a functional specification against which they had to deliver. There is also a broader observation here. While import substitution of components does not necessarily lead to higher performance, it is often accompanied by lower-level process innovation that allows the supplier firm(s) to manufacture components more efficiently. Also shared by the two cases was the impetus arising from the desire to demonstrate green themes in major events and broadly similar governance. Looking at the differences, we may note sensitivity to the timing of the procurement, with Shenzhen able to pursue a more technologically ambitious route. This also reflected a major disparity in the strength of the industrial bases of the two cities. Although different in nature, both cases relied on close and long-standing relationships between buyers and suppliers. As the institutional infrastructure is largely not formalized, interpersonal relationships play a critical role in implementing policies in China, especially in regions. In the absence of a well-regulated competition environment, the best strategy for a company (which is normally the supplier for the public sector) to adopt is to maintain

Table 7.3 Cross-case analysis: identified issues

	Jinan	Shenzhen
Timing	One of the earliest procurements of NEV in China; high price; cautious move; no charging infrastructure so hybrid NEVs were chosen	Midterm of the NEV demonstration program; lowered price; ambitious move attempting to build infrastructure and use diversified types of NEVs
Driving force	Driven by 'green themes' in large-scale events; other examples include Beijing Olympics 2008, Shanghai EXPO 2010, Guangzhou Asian Games 2010	
Local NEV industry status	Relatively weak industrial base according to the criteria set by the central government; trying to nurture local industry by using local market to attract NEV manufacturers	Very strong industrial base; several local NEV manufacturers which are leaders in the country; trying to strengthen the local new energy sector through large-scale NEV procurement
Budget and quantity	100 million yuan for the first 100, and 90 million yuan for the second 100 hybrid coaches	Undisclosed budget; 2011 NEVs based on various technologies (for example, pure electric, hybrid, and fuel cell) were procured
Governance set-up	Cross-departmental administrative group to coordinate the project; supervised by a cross-ministry leading group	
Tendering procedure	Open tendering organized by the provincial government procurement centre; designs were evaluated rather than actual products	Restricted tendering organized by the city government; invitations to leading suppliers in the country were sent out and alternative designs were compared
Interaction mode between stakeholders	Long-term cooperation and trust between the operating company and the supplier company	Pilot projects and regular communications between the city government and the suppliers
Type of innovation	Incremental; 'new to the country' or 'new to the region'; catching up with world-leading technologies	
Outcomes and impacts	Improvement and maturation of the product; lowering of the price; moderate improvement of the local transport system; awareness-raising; hesitation about next moves	Improvement of product design, public infrastructure and the local transport system; contributed to local new energy industry and related standardization; diffusion to other regions

sufficient communication and interactions with key consumers and the government.

Within their own limited terms, both cases could be deemed to have been a success in that products were developed and came into use to socially beneficial effect. However, a broader economic perspective raises some fundamental questions. Most important of these is the apparent role of regional protectionism. In these and other cities, governments were as much motivated by establishing or strengthening an NEV industry in their region as by the benefits to the procurer. This in itself does not have to be a problem; it is one of the principles behind PPI. However, in these cases the results seem to have been achieved by suppressing competition or at least restricting it to local favoured suppliers. The effect of this is restriction of not only price competition but also technological competition. It can be predicted that this will result in wasteful overcapacity in China, and probably a delay of some years before the wider forces of competition and agglomeration rationalize the sector. Furthermore, by segmenting the market in this way, the incentives for each innovating firm are substantially reduced.

The setting of quantitative policy targets is also questionable. How should policy goals be modified to suit social challenges and promote an innovation orientation? What criteria should be used to measure the success of policy implementation? The central government set up the demonstration program by selecting a range of promising cities and allocating them certain quantitative targets to achieve; in terms of evaluation, the government introduced an elimination strategy to screen out laggards by the end of the third year. This policy measure is effective in terms of stimulating large-scale commercialization effort and raising awareness in a short time, but meanwhile it induced inter-regional competition and aggravated protectionism, which further leads to overcommitment of development goals and a danger of sightless and low-quality industry expansion. The competition between administrations cannot simply be reduced to economic rationality. There is a strong suggestion that 'face' and prestige are associated with activity in this sector, even to the extent that it has been called 'NEV fever'.

Chinese policy implementation mechanisms have sometimes been efficient and sometimes not. On the one hand, due to the strictly designed center-locality institutional structure of the political system and the top-down nature of most policies, implementation processes in many cases appear to be much faster than in Western countries. Regions are normally competitors with each other. They respond to the stimulus of central government by designing coherent regional policies rapidly. If the policies are successful, regions can benefit their reputation with central government. However, as regions respond to central government policy, they may find themselves in conflict with their own local stakeholders. One consequence of this is that the targets and outcomes of policy implementation, which are set and evaluated quantitatively, are incapable of reflecting the actual development of the subjects which the policies are initially targeting. The Chinese have a saying to satirize this phenomenon, 'When the upper level government has a measure, the lower level government has a countermeasure', which reflects the exact fact regarding policy implementation in China. This is the resistance side.

The cases are also interesting in that they extend to efforts to catalyze private procurement of socially desirable goods (that is, greener vehicles). The issue here is the small number of cities where subsidies for private buyers were available. Due to the size of the country and uneven status of regional development, pilot programs are frequently adopted by Chinese governments to test policies. Early examples of this approach were the special economic zones introduced by Deng Xiaoping. Today, the selection of pilot regions is more critical and challenging as regions differ in a more complicated way. Major distinctions lie not only in economic status, but also in industry structure, market demand and competitive advantage. In this NEV case, the policy for subsidizing the private consumer only targets six cities in the country. This restrains the willingness for private consumption in other localities. For instance in Jinan, according to interviewee R4O_NEV, some citizens are interested in NEVs, but are hesitating to buy because Jinan is not one of the demonstration cities for private consumer subsidies, and NEVs at their original price are far more expensive than traditional cars.

To conclude: in many ways the story of NEVs has been one of attempting to create a lead market. The problem is that only a narrow range of instruments were used to do this: a highly targeted procurement process, a limited provision of charging infrastructure and some direct subsidies (leaving aside the complex combinations of local and national support). A true lead market would be based on provision of a wider range of favorable framework conditions, for example in the domain of regulations and standards. Part of these framework conditions should be a fully competitive procurement framework, albeit one that maintains the flow of information between supplier and purchaser. Competition between authorities to provide the right conditions for firms is a positive force, but attempts to tilt the market in favor of particular companies can lead only to inefficiency and suboptimal technological development.

NOTES

1. This definition is provided by the State Council in the NEV Industry Development Plan (2012–2020), see http://www.gov.cn/zwgk/2012-07/09/content_2179032.htm (accessed February 19, 2013).
2. The program is also called 'Ten Cities, Thousands of NEVs' (*Shi Cheng Qian Liang*) by practitioners in China.
3. See http://www.most.gov.cn/ztzl/kjzjbxsh/kjzjbxmt/201302/t20130204_99599.htm (accessed February 19, 2013).
4. NEV procurements covered by media include Beijing: http://auto.qq.com/a/20130103/000017.htm;Changchun:http://www.d1ev.com/news-7625/; Hefei: http://news.hf365.com/system/2012/11/22/012755689.shtml (accessed February 19, 2013).
5. See http://www.jinan.gov.cn (accessed May 31, 2013).
6. See GDP ranking of Chinese cities: http://wenwen.soso.com/z/q278620515.htm (accessed February 19, 2013).
7. See http://www.most.gov.cn/kjbgz/200905/t20090527_69944.htm (accessed February 19, 2013).
8. See http://www.nytimes.com/2012/04/20/business/global/rural-chinese-flock-to-tiny-electric-cars.html?_r=0 (accessed May 31, 2013).
9. For example, Geely Co. Ltd, http://www.geely.com.sa/ec8-factory.html (accessed May 31, 2013).
10. See http://english.sz.gov.cn (accessed May 31, 2013).
11. See ranking of the top 25 most innovative cities in mainland China by Forbes, http://www.china.org.cn/top10/2011–10/25/content_23721744.htm (accessed May 31, 2013).
12. BYD website: http://www.byd.com.cn/views/home/indexe.htm (accessed May 31, 2013).
13. Wuzhoulong website: http://www.wzlmotors.com/en/ (accessed May 31, 2013).
14. Information for this model http://en.wikipedia.org/wiki/BYD_F3DM (accessed May 31, 2013).
15. For information on the E6 model see http://en.wikipedia.org/wiki/BYD_e6 (accessed May 31, 2013).
16. 'Report on the development indicators of NEV demonstration program participant cities (2011)', http://www.evtimes.cn/plus/view-32531.html (accessed May 31, 2013).

REFERENCES

Centre for Strategy and Evaluation Services (CSES) and Oxford Research (2011). *Final Evaluation of the Lead Market Initiative Final Report*. http://ec.europa.eu/enterprise/policies/innovation/policy/lead-market-initiative/files/final-eval-lmi_en.pdf (accessed July 20, 2013).

China Automotive Technology and Research Centre (CATARC) (2010). *The 2010 Yearbook on New Energy Vehicles* (in Chinese). Beijing: China Economic Publishing House.

Edler, J. and Georghiou, L. (2007). Public procurement and innovation: resurrecting the demand side. *Research Policy*, 36(7), 949–963.

EUCCC (2011). *Public Procurement in China: European Business Experiences Competing for Public Contracts in China*. Beijing. http://www.publictendering.com/pdf/PPStudyENFinal.pdf (accessed August 15, 2013).

Gong, H., Wang, M.Q. and Wang, H. (2013). New energy vehicles in China: policies, demonstration, and progress. *Mitigation and Adaptation Strategies for Global Change*, 18(2), 207–228.

Guangdong Provincial Government (2009). *Plan on Adjustment and Revitalisation of Automobile Industry in Guangdong Province (2009–2015) (in Chinese)*. Policy code Yuefu[2009]77. Guangzhou. http://www.gdei.gov.cn/zwgk/jmzk/gdjm/200910/200910/t20091023_80476.html (accessed August 18, 2013).

Guangdong Provincial Government (2010). *Guangdong Action Plan on Electric Vehicles* (in Chinese). Policy code Yuefuhan[2010]50. Guangzhou. http://zwgk.gd.gov.cn/006939748/201003/t20100324_11790.html (accessed August 18, 2013).

Huang, S., Wu, Z. and Hong, L. (2012). *The 2011 Annual Report on New Energy Coaches Demonstration Programme in China* (in Chinese). Beijing. http://www.d1ev.com/news-10333/ (accessed August 17, 2013).

Jinan City Government (2009a). *Plan on Adjusting and Revitalizing Automobile Industry in Jinan City (2009–2011)* (in Chinese). Jinan. http://www.jinan.gov.cn/art/2009/7/7/art_76_190243.html (accessed July 17, 2013).

Jinan City Government (2009b). *Plan on Development of New Energy Industry in Jinan (2009–2011)* (in Chinese). Jinan. http://www.jinan.gov.cn/art/2009/7/13/art_76_190765.html (accessed July 17, 2013).

Li, Y. (2011). Public procurement as a demand-side innovation policy tool in China – a national level case study. DRUID Summer Conference. Copenhagen.

Li, Y. (2013). Public procurement as a demand-side innovation policy in China – an exploratory and evaluative study. PhD thesis, University of Manchester.

Li, Y. and Georghiou, L. (2014). Accrediting and signalling new technology – use of procurement for innovation in China. Work in progress.

Lin, J. and Wang, X. (2012). A research on automobile industry innovation capability in China. *IEEE International Conference on Computer Science and Automation Engineering (CSAE)*. Zhangjiajie: IEEE, pp. 600–603.

Liu, Y. and Kokko, A. (2013). Who does what in China's new energy vehicle industry? *Energy Policy*, 57, 21–29.

MOF and MOST (2009). *Circular on Conducting Pilot Projects for Demonstration and Promotion of NEVs (in Chinese)*. Policy code Caijian[2009]6. Beijing. http://jjs.mof.gov.cn/zhengwuxinxi/zhengcefagui/200902/t20090205_111617.html (accessed August 18, 2013).

MOF, MOST, MIIT and NDRC (2010a). *Circular on Conducting Pilot Projects for Private Consumer Subsidies of New Engery Vehicles* (in Chinese). Policy code Caijian[2010]230. Beijing. http://www.sdpc.gov.cn/zcfb/zcfbqt/2010qt/t20100603_351147.htm (accessed August 17, 2013).

MOF, MOST, MIIT and NDRC (2010b). *Circular on Increasing the Number of Cities for Demonstration and Promotion of New Energy Vehicles in the Public Section* (in Chinese). Policy code Caijian[2010]434. Beijing. http://www.jscz.gov.cn/pub/jscz/zfxxgk/zfxxgkml/ywgz/dwgnywwj/201104/t20110422_20461.html (accessed August 18, 2013).

MOF, MOST, MIIT and NDRC (2011). *Circular on Continuing Conducting Pilot Projects for Demonstration and Promotion of NEVs*. Policy code Caibanjian[2011]149. Beijing.

O'Brien, R.D. (2010). China's indigenous innovation – origins, components and ramifications. *China Security*, 6(3), 51–65.

Shandong Provincial Government (2009a). *Administrative Measures on Financial Support of NEV Demonstration Programme in Shandong (Trial)* (in Chinese). Policy code Luzhengbanfa [2009]130. Jinan. http://www2.shandong.gov.cn/art/2009/12/24/art_3883_1406.html (accessed August 18, 2013).

Shandong Provincial Government (2009b). *Measures on Accelerating the Development of NEV Industry in Shandong (in Chinese)*. Policy code Luzhengbanfa[2009]64. Jinan. http://www.shandong.gov.cn/art/2009/9/7/art_956_3181.html (accessed August 18, 2013).

Shandong Provincial Government (2009c). *Measures on Accelerating the Development of New Energy Industry in Shandong* (in Chinese). Policy code Luzhengfa[2009]140. Jinan. http://www.shandong.gov.cn/art/2009/12/24/art_956_3622.html (accessed August 16, 2013).

Shandong Provincial Government (2009d). *Plan on Adjusting and Revitalizing Automobile Industry in Shandong (2009–2011)* (in Chinese). Policy code Luzhengfa[2009]44. Jinan. http://www2.shandong.gov.cn/art/2009/4/25/art_3883_1531.html (accessed August 16, 2013).

Shenzhen City Government (2009a). *Implementation Measures of Building Public Charging Infrastructure for NEVs in Shenzhen (2009–2012)* (in

Chinese). Shenzhen. http://paper.ce.cn/jjrb/html/2012-06/12/content_214246.htm (accessed August 19, 2013).
Shenzhen City Government (2009b). *Plan on Revitalizing and Developing the New Energy Industry in Shenzhen (2009–2015)* (in Chinese). Policy code Shenfu[2009]240. Shenzhen. http://www.sz.gov.cn/zfgb/2010/gb681/201001/t20100112_1424099.htm (accessed August 18, 2013).
Shenzhen City Government (2011). *The 12th Five-year Plan on NEV Industry Base in Shenzhen* (in Chinese). Shenzhen. http://news.xinhuanet.com/2010-10/22/c_12688380.htm (accessed August 19, 2013).
State Council (2011a). *China's Policies and Actions for Addressing Climate Change*. Beijing: Foreign Languages Press. Beijing. http://www.gov.cn/english/official/2011-11/22/content_2000272.htm (accessed August 17, 2013).
State Council (2011b). General implementation plan for energy saving and carbon emission reduction during the 'twelfth five-year plan' period. Guofa[2011]26.
State Council (2012). Development plan for the energy saving and new energy vehicles industry (2012–2020). Guofa [2012]22. http://www.gov.cn/zwgk/2012-07/09/content_2179032.htm.
Sun, L. (2012). Development and policies of new energy vehicles in China. *Asian Social Science*, 8(2), 86–94.
USCBC (2010). Issue brief: new developments in China's domestic innovation and procurement policies.
USCBC (2011a). China's Innovation and Government Procurement Policies: Next Steps.
USCBC (2011b). PRC government actions to meet bilateral commitments on indigenous innovation and government procurement.
Wang, P. (2009). China's accession to the WTO Government Procurement Agreement – challenges and the way forward. *Journal of International Economic Law*, 12(3), 663–706.
World Bank (2011). *The China New Energy Vehicles Programme – Challenges and Opportunities*. Washington, DC. http://documents.worldbank.org/curated/en/2011/04/14082658/china-new-energy-vehicles-programme-challenges-opportunities (accessed August 15, 2013).

8. Public procurement for e-government services: challenges and problems related to the implementation of a new innovative scheme in Greek local authorities

Yannis Caloghirou, Aimilia Protogerou and Panagiotis Panaghiotopoulos

INTRODUCTION

This chapter provides empirical evidence on the role of public procurement for innovation (PPI) in the field of e-government. It does so by examining different aspects of a pioneer public procurement practice for the pilot provision of local e-government services in Greece.

In particular, the study focuses on the Local Government Application Framework (LGAF) pilot project that was launched by the Central Union of the Greek Municipalities (KEDE) in 2007. The objective of the project was the development of a centralized software system (platform) offering high-quality e-government services to citizens and local businesses. The project was implemented in two stages: the first focused on the design, development and delivery of the platform; and the second involved the pilot use of the platform by selected local authorities.

The empirical part of the chapter is based on case study work[1] using a semi-structured questionnaire. A series of in-depth interviews with the key actors involved in the platform's development sheds light on: (1) the objective of the LGAF project; (2) its innovative characteristics; (3) the needs addressed; (4) the stages and the outcome of the procurement process; (5) the obstacles to success; and (6) its long-term potential benefits.

Our findings suggest that the LGAF project can be characterized as a PPI practice with aspects of both direct and cooperative and catalytic procurement. This stems from the fact that a central public authority (KEDE) organized and coordinated a project aiming at the development of a system that would be used by other peripheral public authorities (municipalities) under the procurement scheme 'design and build once and apply many times'. This project attempted to address the general challenge of providing value-added e-government services, creating at the same time more efficient internal management structures and achieving significant scale economies. In the Greek context this new procurement approach could be considered as an effort to find substitutes for the usual inefficient and ineffective practices whereby each local authority buys an almost identical information and communication technology (ICT) package, paying separate licence fees. It is worth noting that these packages are usually turnkey solutions, that is, they are not products tailored to the specific requirements of each municipality, but at best a limited array of ready-made services.

The expected benefits that guided KEDE's decision towards this procurement were: (1) the delivery of high-quality e-government services to citizens and local businesses, taking full advantage of the rapid development and spread of new advanced technologies; and (2) the avoidance of possible future costs, caused by inflexibility due to vendor lock-ins and the high price of new development, by improving existing systems in new ways.

A series of issues that came up during the implementation phase resulted in the delayed development and delivery of the platform. Most strikingly, its actual operation is still pending; that is, the LGAF platform is still not in use. The considerable delay in the system's development can be primarily related to its redesign shortly after the project's initiation. The redesign decision was realized through the adoption of a state-of-the-art technological solution based on a flexible architecture that could meet the project's functional requirements in a more effective and efficient way. In addition, several issues such as the insufficient expertise of the purchasing agency; the relatively limited involvement of the end-users; the inefficient technical, risk and relationship procurement management; the legal framework of the project; and the weaknesses of the Greek ICT ecosystem constituted further significant impediments to the successful overall realization of the project. The utilization of the final product requires additional funding and extensive organizational work in terms of redesigning and integrating the municipalities' internal processes. Thus, its main added value lies in its potential long-term benefits. In particular, the platform's design characteristics render feasible its future reuse and transfer to other public and private national entities. Even more, its architectural features support its interconnection and interoperability with other information systems of the public sector in Greece and other European Union (EU) countries, so that cross-sectoral and cross-border services could be provided in the future.

The chapter continues with a brief presentation of the provision of e-government services in Europe. Then a short description of the procurement framework in Greece follows with emphasis on the e-government services offered at the local level. The fourth section presents the LGAF case study and discusses the empirical findings, focusing mainly on the stages and the result of the procurement process and the major obstacles to the project's successful implementation. The chapter ends with a number of conclusions regarding the type and characteristics of the specific PPI practice, its long-term potential, and the challenges associated with the management of the PPI process.

THE CONTEXT OF E-GOVERNMENT IN EUROPE

E-government (electronic government) refers to the utilization of ICTs, and other web-based telecommunication technologies, to improve the efficiency and effectiveness of service delivery in the public sector (Jeong, 2007).

E-government changes considerably the way public services are delivered and generally the way in which government interacts with citizens and businesses. Thus, it can be considered as a field that provides significant room for public procurement for innovation (PPI). The benefits resulting from an extensive realization of e-government concern a large variety of actors. First, e-government can enhance the public sector's productivity, increase transparency, and lead, in consequence, to less corruption, cost reductions and increased public revenue. At the same time, it can result in better delivery of public services to citizens by ensuring time and cost savings and generally by upgrading their quality of life. Furthermore, e-government can improve the interactions of government with industry, strengthening in this way the private sector's productivity and competitiveness prospects.

The EU's e-government services are becoming increasingly interactive and transactional, while the quality of service delivery has been significantly improved over time (European Commission, 2013). However, three key messages emerge from a recent e-government Benchmark Report (European Commission, 2013, p. 60). First, there is an increasing challenge to meet citizens' ever-growing expectations for simple, easily accessible and swift public services, especially when taking into account what modern technologies can do. It appears, therefore, that the shift in e-government thinking towards designing more user-centric services is not yet fully embraced in Europe. Second, governments are not fully reaping the potential benefits of e-government. At present, e-government use in the EU is on average at 46 per cent. In addition, there is slow progress in usage considering the number of fully available online services. This causes the required investments in ICT for public service provision to be underutilized and thus inefficient in economic terms (for example, the expected cost savings are partially realized), raising the question of how effectively European countries are getting citizens to use online services.

Finally, transformation is needed to accomplish a new generation of e-government services focusing their design primarily on an 'outside-in' approach; that is, design through the eyes of the user (citizens and businesses) rather than through the eyes of the provider. The need for adopting novel ways in the design and delivery of services is based on motives such as economic recovery, severe budget constraints and customer expectations. Public administrations should continuously adapt to rapidly shifting environments and, thus, flexible and fully interoperable organizations and systems are required.

Interoperability[2] is considered as a crucial factor to address the above-mentioned challenges, and the emphasis that the European Commission puts on this issue is very strong. In particular, the Digital Agenda for Europe[3] (European Commission, 2010a) sets out a common and coherent approach to interoperability as well as priorities for actions. Moreover, it introduces a conceptual model (Figure 8.1) for developing European cross-border and cross-sectoral[4] life-event e-government services.[5] It presents a building block approach to construct these, allowing

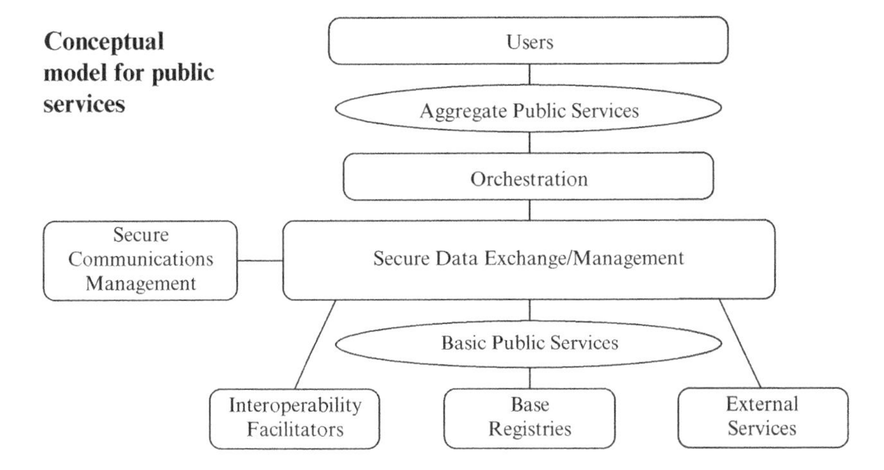

Source: European Commission (2010c, p. 14).

Figure 8.1 The conceptual model for public services of the European Interoperability Framework

service components to be interconnected, and promotes the reuse of information, concepts, patterns, solutions and specifications in member states and at the European level. The model can be subdivided into three main layers: basic public services, secure data exchange and management, and aggregate public services (European Commission, 2010c).

The lowest layer of the model deals with the most basic service components from which the European public services can be built. The most important components are base registries which provide reliable sources of basic information on people, companies, vehicles, licences, buildings, locations and roads. These registries are under the legal control of public administrations and are maintained by them, but the relevant information should be made available for wider reuse with the appropriate security and privacy measures. Interoperability facilitators provide services such as translation of protocols, formats and languages, or act as information brokers. The third type of basic service components are services provided by external parties, such as payment services provided by financial institutions or connectivity services provided by telecommunications providers. The secure data exchange and management layer is central to the conceptual model since all access to basic public services goes through it. This layer should allow the secure exchange of certified messages, records, forms and other kinds of information between the different systems. In addition, it should also handle specific security requirements such as electronic signatures, certification, encryption and so on. Aggregate public services, depicted in the upper layer of Figure 8.1, are constructed by grouping basic public services that can be accessed in a secure and controlled way. They can be provided by several administrations at any level: local, regional, national or even EU level. A typical aggregate service should appear to its users (administrations, businesses or citizens) as a single service. However, behind the scenes, transactions may be implemented across borders, sectors and administrative levels (European Commission, 2010c).

There are multiple significant benefits of achieving interoperability in an e-government context. From the standpoint of public services, interoperability allows for better coordination of government agency programmes and services, improved service delivery and access, and more efficient technology management and maintenance. Furthermore, public administration can avoid potential future costs such as reliance on a single vendor and the high price of new development by leveraging existing systems in new ways. From the standpoint of policy-makers, e-government interoperability can improve data-gathering and parsing techniques, resulting in more efficient and accurate decision-making. Finally, it can also promote international cooperation by assisting in the creation of infrastructure to address cross-border issues and to allow for the provision of cross-border services, that is, services that require the coordination of public authorities in different member states, an important pillar for the realization of a European single market (UNDP, 2007; Novakouski and Lewis, 2012).

Currently, there are some important large-scale pilot projects at the EU level following the aforementioned conceptual model. Furthermore, some European countries have developed strategies and initiatives for the collaboration of different government entities and levels in order to coordinate e-government activities (European Commission, 2010d).

Summing up, e-government in Europe can be realized at three levels. The first one refers to interactive and transactional basic services at a national level; the second concerns building interoperable life-event services at a national level; and the third regards cross-border life-event services at an EU level by enhancing interoperability of public authorities in different member states. While the first e-government level appears to be adequately implemented, the realization of the next two, which implies much higher added value for citizens and businesses and the public administration, is still far from satisfactory.

PUBLIC PROCUREMENT FOR E-GOVERNMENT SERVICES IN GREECE

In Greece, the use of public procurement to stimulate demand for innovative products or services is fragmented and based on priorities and policies set by various ministries (Nioras, 2011). Although Nioras (2011) reports a shift towards more emphasis on demand-side measures supporting innovation, cost-effectiveness and rationalization tend to be the dominant rules for any type of public spending (European Commission, 2012). Political hostilities in the past have created a very unfavourable climate for trust, an important prerequisite for PPI when criteria other than price are to play a critical role in the evaluation procedure. The lack of adequate human resources and skills on the buyer's side, along with the bureaucracy governing the operation of the public sector, may be considered as additional inhibitory factors (Edler et al., 2005). Hence, in general, the Greek context does not seem to favour PPI, and so far there is little evidence of increasing activities in this direction. A characteristic case is procurement of military equipment. Greece has one of the highest levels of defence expenditure as a share of gross domestic product (GDP) among the EU and North Atlantic Treaty Organization (NATO) countries. Nevertheless, the Ministry of Defence's research and development (R&D) expenses were less than 1 per cent of total government appropriations for R&D (European Commission, 2012).

An exception to the general implementation rules of the public procurement framework is the procurement of ICTs, co-financed by the EU in the context of the Community Support Framework. In this case, specific procedures are foreseen to ensure that the cost-effective spending is accompanied by compliance with the technological developments (Edler et al., 2005).

However, a recent study (SEV/IOBE/NTUA, 2011) indicates that, in general, the Greek public sector cannot be considered as an intelligent buyer of ICT and e-government solutions. What is more, the usage of ICTs at all government levels (national, regional and local) is far from satisfactory, and the development and use of sophisticated e-government services are considerably low. Therefore, the public sector has not been a stimulating factor for innovation in the ICT sector. This can be attributed to various factors such as insufficiency of human capital, resistance behaviours and other institutional aspects (SEV/IOBE/NTUA, 2011).

In particular, a large share of the public procurement process is taking place in the context of co-funded EU–national projects due to the financial opportunities that they offer. Yet, in the majority of cases, public authorities do not have the required human resources, knowledge and organizational capabilities for an efficient involvement in such projects; that is, they are not qualified to specify the functional and quality requirements, assess the different tenders, tightly monitor and evaluate the implementation progress and, finally, test and accept the delivered product or service. Moreover, by using strict – sometimes outdated and very restrictive – technical specifications rather than functional requirements in the relative calls, public authorities do not encourage the development of innovative products or services (SEV et al., 2011).

In addition, the legal framework favours large and well-established firms as prime contractors due to their previous experience and financial credibility, which are considered as two basic award criteria. This fact limits considerably the participation of younger, smaller and would-be innovative ICT firms. Furthermore, in cases of public cooperative and catalytic procurement (for example, municipalities as end-users) (Edler and Georghiou, 2007; Edquist and Zabala-Iturriagagoitia, 2012), Greek central public agencies do not usually develop complementary instruments (for example, improve awareness and training of the members of the municipalities' councils in ICTs) to ensure adequate involvement of end-users (SEV et al., 2011). Therefore, the specific weaknesses on the public demand side, and the legal framework of the EU co-funded ICT projects, can be assumed to be barriers to creating innovative e-government services (SEV et al., 2011).

The same study also suggests that the supply side does not support innovation since firms active in the Greek information technology (IT) industry are primarily offering products of low sophistication, specialization and standardization. This can be partly attributed to the low quality of public demand, but it can also be related to private demand. The private sector's demand is primarily determined by the unconsolidated structure of the Greek economy, that is, the relatively large number of small and especially micro firms that typically do not demand high-level IT solutions (SEV et al., 2011).

Therefore, the majority of firms in the IT sector appear to have conceived of the private and public demand for computer and e-government services not as an opportunity to enhance their stock of knowledge and achieve sustainable competitive advantage, but only as a means of temporarily increasing their profits. Due to this short-term business strategy, they are experiencing quite intensely the negative impacts of the current economic crisis (SEV et al., 2011).

Greek Local Authorities and E-Government Services

The weakness of public demand in procuring innovative ICT products and services also seems to hold at the level of Greek local authorities. This shortcoming was confirmed by a recent large-scale survey[6] (KEDE and NTUA, 2011), which examined the extent of ICT utilization in Greek municipalities and their procurement practices and capabilities in designing, undertaking and operating ICT projects.

The empirical findings (KEDE and NTUA, 2011) resulting from this survey suggest that the majority of the surveyed municipalities have not developed an explicit ICT strategy. Furthermore, over half of the municipalities do not have a formal IT unit, and even in cases where such units exist, the number of employees appears not to be adequate to support their operation.

Most importantly, there seems to be a lack of human resources and organizational capabilities for carrying out PPI practices in the field of ICT. In particular, only one out of ten municipalities appears to have highly skilled personnel for the design of ICT proposals, while the relevant shares for tender assessment and project monitoring are 34 per cent and 23 per cent, respectively. Moreover, only a small share of the surveyed local authorities uses internal and external (with citizens and local businesses) consultation on a regular basis in order to prioritize ICT projects based on actual needs. Also, only in one-fifth of the municipalities do the administrative units, which are also users of a project's output, participate actively in setting out the functional requirements. Finally, only 15 per cent dedicate adequate financial resources to the technical maintenance and long-term upgrading of ICT projects. Regarding the issue of user skills, the majority of municipalities appear not to have employees with sufficient digital skills to operate the municipality's software systems. Moreover, only a small share of them (12 per cent) undertake training initiatives to enhance the digital skills of their employees and councillors (KEDE and NTUA, 2011).

The aforementioned weaknesses result in the low quality and interactivity of services offered to citizens and local businesses. Especially, more than half of the municipalities do not provide e-government services at all, while the rest of them give mainly information-based e-government services. More interactive and transactional services are offered only by a small share of large and medium-sized municipalities. Accordingly, the demand for these services remains extremely low.

By and large, this research work points out that Greek municipalities do not have the strategy, human capital and internal capabilities required to be intelligent customers and users of IT and e-government solutions. Nevertheless, it must be noted that there is a relative small number of pioneer local authorities that appear to have the resources and competences needed to improve and further develop their e-government services.

THE LOCAL GOVERNMENT APPLICATION FRAMEWORK PROJECT

The Local Government Application Framework (LGAF) pilot project was launched by the Central Union of Greek Municipalities (KEDE)[7] in 2007 as a part of its ICT strategic plan. It was a co-funded EU–national project with a total budget of €1.6 million aiming at the development of a centralized software system (platform), and at the interconnection and interoperability of this system with the existing (legacy) application systems of local authorities. The ultimate goal of the project was the pilot offering of high-quality e-government services to citizens and local businesses in eight municipalities of various sizes and geographical characteristics.

The carrying out of the work involved two basic stages:

1. The design and development of the platform and its delivery as a product to the purchasing agency (KEDE).
2. The platform's utilization by end-users for the pilot delivery of e-government services.

The case study work was conducted using a semi-structured questionnaire focusing on different aspects of the specific public procurement process, such as: (1) the targeting of the LGAF project; (2) the stages and the result of the procurement process; (3) the obstacles to success; (4) the project's characteristics as a PPI practice; (5) the importance of specific design and procurement management capabilities; and (6) the long-term potential of the project. The case study protocol guided a series of in-depth interviews with the founders or chief executive officers (CEOs) of the firms and the research centre officials who played a key role in the design and execution of the project. The interviews were supplemented with extensive reviews of the project's documentation.

The Stages of the Procurement Process

In 2005, KEDE decided to exploit the financial opportunity offered by the EU Cohesion Policy Fund to carry out a pilot project with dual targeting: (1) the delivery of value-added online services to citizens and local businesses; and (2) a more efficient management of local authorities' resources and organizational processes. This targeting was translated into the following design principles and functional requirements:

1. The development of a centralized platform using the 'build once, use many' strategy, which is a more efficient approach for government, rather than each agency spending time and resources to build its own e-government tools and systems (Gallerani-Cineca, 2011).
2. The adoption of open standards[8] to ensure the platform's technical interoperability[9] with the municipalities' existing (legacy) systems, and at the same time promote the scalability[10] and reusability[11] of the platform and its potential interoperability with other information systems of the public administration.
3. The implementation of this platform by using open source software (OSS), that is, software whose source is available

> for free and can be copied, modified and expanded by any developer.[12] The use of OSS allows the avoidance of future costs such as the obligation of licence fee payment and inflexibility due to vendor lock-in.[13] Furthermore, it facilitates the transfer and use of the produced technological knowledge by other public authorities.

The procurement method for acquiring the platform was an open call which included the aforementioned requirements and suggested a number of OSS packages already in use by local governments in other European countries (the UK, Sweden, and so on). The contract was awarded to a large, well-established Greek IT firm that decided to use one of the proposed OSS solutions (APLAWS) as the base for the new system's development.

APLAWS (Accessible and Personalised Local Authority Website System) was a nationally funded scheme designed to develop technology that could be reused across the UK. It was launched in 2001, and was a partnership of five London boroughs, the Greater London Authority, three charities and three commercial companies. The overall aim of the project was to deliver a freely available content management solution that met the interoperability, metadata and accessibility requirements set for UK local authorities. In 2003, following the success of the first phase, the APLAWS Project was incorporated into the Local Authority Websites National Project, and in the period 2003–2005 the base system and the various website extensions were improved considerably. In particular, the content management system was further enhanced. This included adding support to use a fully open source stack and separating the various components of the content management system to ensure they can be developed and extended independently. In 2006, more than 30 UK local authorities and numerous international organizations (United Nations Development Programme, Malaysian and German universities, Brisbane local authority in Australia and the Chinese meteorological bureau) were using it for either their website or their intranet.[14]

In 2007, KEDE and the prime contractor decided to redesign the project towards a more state-of-the-art technological solution[15] that could meet more effectively the prescribed functional requirements, and especially enhance the interoperability, scalability and reusability of the platform. In particular, the main argument for the project's technical reorientation was that although APLAWS was a pioneer project at the time it was developed, it could not be easily adapted to the new technological context of the rapidly evolving web-service technologies.

Web-service technologies support the adoption of more flexible system architectures such as service-oriented architecture (SOA). SOA is an enterprise-wide IT architecture that offers a better way of designing integrable, reusable application assets, orchestrated from existing services rather than rebuilt from scratch.[16] Its main characteristic is its implementation of a service platform consisting of many services that signify elements of business processes that can be combined and recombined into different solutions and scenarios, as determined by the business needs (Biebertein et al., 2006). This capability provides organizations with the flexibility needed to respond quickly and effectively to new situations and requirements (UNDP, 2007). Consequently, SOA is considered as the best underlying paradigm with which to begin to roll out cross-agency and cross-border e-government services,[17] and is proposed as an implementation of the building block approach of the European Interoperability Framework.

In order to exploit the technological and operational advantages that the SOA approach offers, KEDE and the prime contractor agreed to build a new and more complex platform consisting of different components.[18] The advantages of the platforms' redesign compared to the initially adopted solution are threefold: (1) higher levels of interoperability with municipalities' existing systems; (2) adaptation and responsiveness to required changes and increased reusability potential; and (3) possibility of interoperation with other information systems of the public administration.

The project's reorientation raised substantially the need for specialized software providers (subcontractors) with expertise in SOA due to the various platform components that had to be developed. Hence, the project team was progressively transformed into a network of specialized service providers (consisting of two research labs and several small and micro firms) which undertook the development of the specific components of this architecture.

The development process of the platform components also demanded an active involvement of the end-users (municipalities), so that their actual needs would be properly addressed. Nevertheless, only one participating municipality contributed to the development of one of the core components of the platform – that is, the business process management system (BPMS) – by re-engineering and modelling the specific organizational processes that would underlie the delivered services.

Furthermore, the project's implementation required a high-level technical management for the efficient coordination of the specialized providers and the integration of the produced components. However, the prime contractor did not seem to be familiar with such a division of innovative labour. The technical management became quickly a non-cooperative bargaining game, where the main project participants pushed towards different technical directions. As a result, the project passed through several transition stages, and the work was delayed considerably.

It should be also noted that, during the implementation phase, some of the subcontractors withdrew from the project due to the inconsistencies of the financial flux[19] and/or their limited capacity to meet the project's technical requirements. However, a larger firm (employing approximately 40 employees) joined the consortium and proved to be very helpful in accelerating the technical work and completing the first basic stage of the project that is, the development of the LGAF platform. Figure 8.2 depicts the main actors pertaining to the demand and the supply side of the project as well as the degree of cooperation among them.

Results

A series of endogenous and systemic obstacles led to delays in the platform delivery and essentially resulted in the partial completion of the project. The first stage of the project, the development and delivery of the LGAF platform to KEDE, was completed with significant delay in December 2011. However, as of November 2013 the second stage has not yet been completed; that is, the LGAF platform has not been put into operation.

Thus, so far the result of the procurement process is the creation of a centralized system that incorporates three core design principles:

1. It adopts the rationale 'build once, use many'.
2. It has been developed by using open standards and open source technologies.
3. It is based on service-oriented architecture.

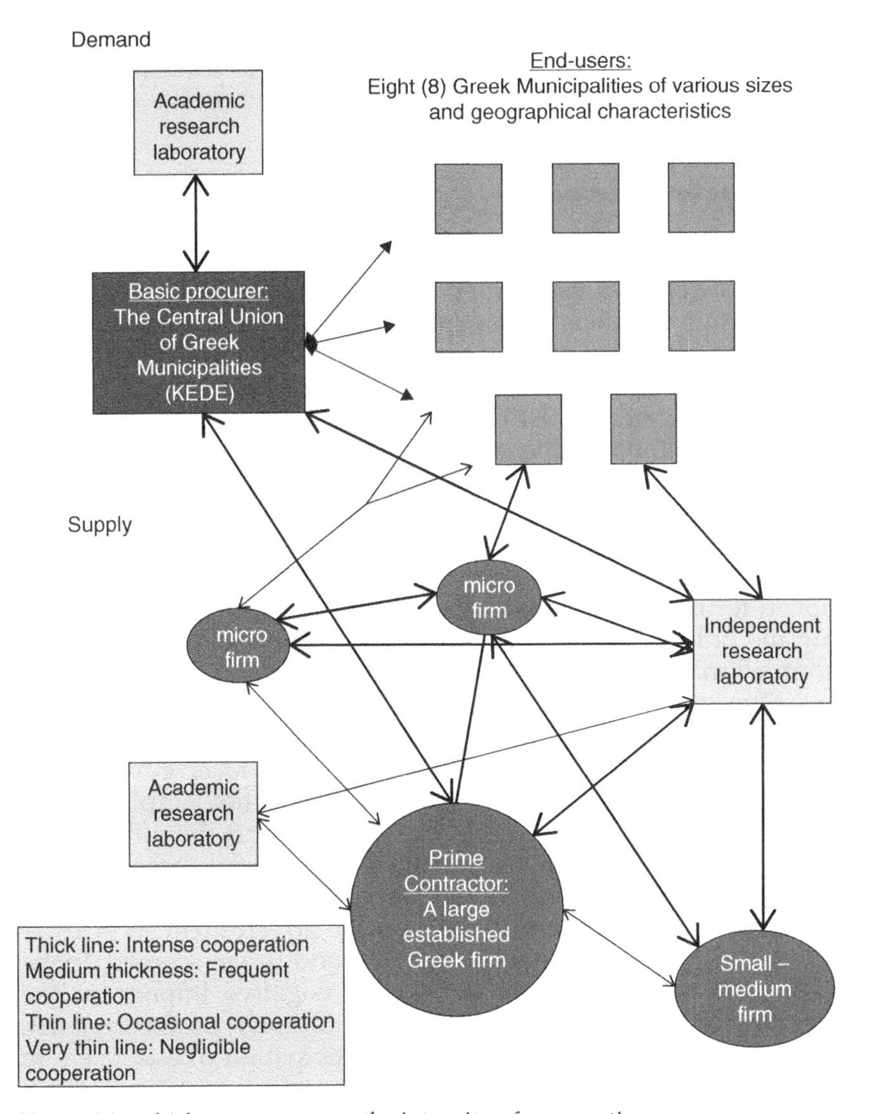

Note: Line thickness represents the intensity of cooperation.

Figure 8.2 The main actors of the LGAF project

Nevertheless, the efficient utilization of the platform further requires: (1) technical interoperability with the municipalities' legacy systems; and (2) preparatory organizational work, namely redesign and modelling of municipalities' internal processes in a uniform way for the production of e-government services. In financial terms, it demands the creation of a business plan that will determine the way in which the operation cost will be covered.

Obstacles to Success

Weaknesses on the side of demand and supply

During the last ten years the basic procurer, KEDE, in extensive consultation with academic experts, has developed certain policy initiatives for ICT adoption and usage at the local government level. As far as the LGAF project is concerned, our empirical findings indicate that KEDE adopted the suggestion of its academic consultant for the procurement of a platform that would provide high-quality e-government services. In addition, KEDE used the technical expertise of its consultant to articulate the basic design principles of the project. However, KEDE did not have the capacity to support this PPI in terms of management, nor in terms of mainstream procurement skills, that is, managing the bidding process, bid evaluation, contract awarding and contract management.

Moreover, KEDE did not use complementary instruments to intensify the engagement of the selected municipalities in the project (for example, increase awareness of local government leaders or enhance training of local authorities employees in ICTs). Such instruments could possibly support the municipalities' active involvement, especially if one takes into account their insufficient technical knowledge, capacity and user skills (see the section on 'Greek local authorities and e-government services').

In addition, the involvement of end-users seemed to be extremely limited, having an overall negative impact on the platform's development and, most importantly, on the system's operation and testing. The limited participation of end-users in the project can be directly related to their selection criteria which were primarily based on size and geography – following the concept of representativeness included in the directives of the 3rd Community Support Framework – and not on their actual interest in the project or their real capacity to get involved in it. A notable exception is a relatively large municipality (Larissa, the capital of Thessaly region) which has participated very actively in the project. This is mainly due to its highly competent chief information officer and experienced IT unit personnel who were able to recognize immediately the project's importance, its potential benefits and the extent of organizational work needed on their part.

Regarding the supply side, our findings suggest that path-dependency in the way IT public procurement projects are designed and carried out in Greece was a significant impediment to the successful completion of the project. The majority of large firms operating in the IT industry are involved in public projects in the traditional way, that is, they are not developing substantially innovative solutions but they mainly offer a limited modification of existing products in response to very restrictive technical specifications (see the section on 'Public procurement for e-government services in Greece'). The prime contractor of the LGAF project belongs to this pool of large firms and as a consequence it appears that it did not have the mindset and skills needed for an active technical and managerial role in such a demanding project. In addition, the Greek market appears to be rather narrow in terms of providing specialized software developers for this type of innovative project. As a consequence, some of the smaller subcontractors failed to meet the project's technical requirements and the consortium delayed to get an efficient form.

The project's redesign

Although the redesign of the LGAF project can be considered as a source of innovation, at the same time it hindered its successful realization. The project incorporated new core design characteristics without making the necessary formal rearrangements that these new features demanded.

In particular, the redesign of the project resulted in assigning part of the contract tasks to subcontractors which were mainly small and young firms. This raised the need for more flexible and steady funding flow to reduce the financial risk of those actors whose involvement in the LGAF project constituted a significant part of their total economic activity. However, the contract designated that the payment of the largest share of the budget would follow the successful closing of the project. As a consequence, serious finance problems were caused to the micro and small partners slowing down at the same time the implementation of the project.

In addition, the contract agreement had initially foreseen that only 10 per cent of the total budget would be dedicated to the initial stage of the LGAF project, the platform development. This is because the main part of the budget would support the participating municipalities in the pilot testing and use of the platform. But although the budget allocation might have been appropriate for the implementation of the platform through the modification of the APLAWS package it proved insufficient for the design and development of a more novel and complex system.

Furthermore, the enhanced innovativeness of the project would be more congruous with the production of a smaller range of e-services for a smaller number of municipalities (maximum three or four instead of eight). In this way, the operation of the platform could be extensively tested and evaluated and the eventual problems could be clearly identified and more easily solved.

CONCLUSIONS

Type and Characteristics of PPI

The LGAF project can be characterized as a PPI practice that attempts to address the challenge of providing upgraded e-government services to citizens and businesses, achieving at the same time a more efficient management of resources and organizational processes and significant economies of scale for local authorities. It has aspects both of direct and of cooperative and catalytic procurement (Edler and Georghiou, 2007; Edquist and Zabala-Iturriagagoitia, 2012) as one central public authority (KEDE) organizes and coordinates the project whose product (platform) will be used mainly by other peripheral public authorities (municipalities).

The innovative character of the LGAF project is closely related to its initial design principles; however, it was further enhanced by taking up the SOA. The adoption of this architecture is actually the main source of technological innovation as it supported the combination of various state-of-the-art or beyond-state-of-the-art technologies for the creation of a new integrated system. This combination process required extensive coordination and resulted in the individual components' enrichment and modification by the specialized developers. As a consequence, the LGAF project can be considered as an adaptive PPI (Edquist and Zabala-Iturriagagoitia, 2012) as it leads to a significant incremental innovation through the integration of various advanced technologies.

It is important to note that as the ultimate objective of this procurement process was the delivery of product-based services (Technopolis, 2011), the created technological innovation can constitute the technical base for extensive organizational innovation. However, this stage of the project remains to be completed and requires the harmonization of the municipalities' internal business processes that will in turn lead to the delivery of high-quality digitized services.

A significant positive side-effect of the LGAF project was that it created opportunities for knowledge-intensive entrepreneurship (Malerba and McKelvey, 2010; Radosevic et al., 2010; Edquist et al., 2010; Timmermans and Zabala-Iturriagagoitia, 2013). The embraced service-oriented architecture resulted in the creation of a complex platform consisting of different components, and provided subcontracting opportunities to specialized software developers. What is more, the use of OSS, due to its positive propensity to knowledge exploration, stimulated even more the participation of small-sized software providers that are highly efficient knowledge absorbers. Hence, the LGAF's innovative characteristics actually supported the creation of a knowledge network among micro and small knowledge-intensive organizations (firms and research and academic institutions) which ultimately led to the system's development. On top of that, the future operation of the platform may create more opportunities for KIE through the support services to the system's end-users.

Long-Term Potential

The long-term potential of the LGAF project – up to now as an experimental public procurement practice – lies in its design features which offer high possibilities of reusability and transferability. This means that the technological knowledge accumulated during the design and development of the platform can be leveraged and also transferred to other government levels (central or national, regional) and functions[20] (economic affairs, health, education, social protection, public order and safety, and so on) leading to proliferative benefits. However, policy and organizational initiatives are needed so that the reuse of the

LGAF platform and the transfer of the accumulated knowledge can take place.

For example, the LGAF platform can be used in the future by a larger group of Greek municipalities. Yet, each municipality that intends to use the platform should carry out the necessary technical and organizational work in order to achieve technical and organizational interoperability with the platform. The LGAF system can also interoperate with public sector information systems in Greece and other EU countries for the delivery of cross-sectoral and cross-border services (see the section on 'The context of e-government in Europe'). Nevertheless, this type of aggregate service requires specific interoperability agreements at different levels (technical, semantic, organizational, legal) between the public authorities involved in the delivery of such services (European Commission, 2010c).

Furthermore, the use of OSS for the development of the platform favours the transfer and utilization of the produced technological knowledge (see the section on 'The stages of the procurement process') by other public organizations (ministries, regional administration, public utilities, and so on) and private organizations (for example, large firms). This demands organizations with sufficient in-house expertise or external service providers in order to be capable to adapt this technological knowledge beneficially to their own needs. Generally, a national policy for the use of OSS by the public (and private) sector could favour the development and transfer of OSS platforms like LGAF between different domains.

Challenges Associated to the Management of the PPI Process

As governments are adopting PPI as a crucial instrument for demand-driven innovation policies, public organizations are facing the need to develop capabilities and skills to manage these novel procurement practices. The innovation perspective sets new challenges for the public agencies and the procurement process itself (Valovirta, 2012).

Although this case study indicates that KEDE has been involved in an innovative public procurement, this central agency and the Greek local authorities in general cannot be characterized as advanced and established buyers of innovative ICT solutions. They also do not seem to have a well-articulated procurement strategy with explicit goals that stimulate innovation.

However, with the first experience at hand, they can gradually build their capacity to support PPI and improve their relevant management capabilities and skills.

PPI can contribute to improved efficiency and effectiveness of public services delivery. However, technological risks may lead to non-completion, underperformance or false performance of procured products or services. It should be acknowledged that risks are inherent in the innovation process and thus can never be fully eliminated. However, managing technology risks that may result in lower than expected performance is a key element in PPI (Valovirta, 2012).

The case study findings suggest that KEDE could not effectively manage the technology risks related to the platform's development, which is as of November 2013 partially completed. More specifically, KEDE was not able to communicate effectively the goals of the procurement among the selected end-users (municipalities) and find ways to increase their engagement in the project's implementation. In any case it should be noted that the end-users' selection was based on the representativeness criteria included in the directives of the project's legal framework.

Knowledge of the market and interaction with suppliers are key management issues. Our case study findings suggest that, once the project's redesign was determined, there was a more or less effective interaction with the market, which led to the enlargement of the project team with small and more innovative subcontractors that were essential to the development of the new technological solution. However, there is a challenge regarding the process of awarding contracts that actually limits the supplier base and leaves public agencies with fewer options to choose from. The current institutional framework, defined by EU directives that try to limit fraud and safeguard delivery of the projects, favours the selection of well-established providers, and at the same time reduces the chances of young, innovative firms without a track record in similar projects to obtain a contract. In this respect, the case study shows that the LGAF project was awarded to a large, well-established Greek firm that most importantly did not have sufficient technical competence to play a critical role in this type of PPI practice.

The relationship and coordination management of the various entities that were responsible for the platform's creation was of high importance due to its increased technical complexity. However, KEDE appeared not to be able to support effectively the different actors' coordination during the procurement process. The prime contractor that was responsible for the partner's technical coordination appeared not to have sufficient technical capacity to undertake this particular new way of developing and implementing an ICT project in the Greek context. As a consequence, the coordination of the partners' work was not as efficient as required by the LGAF-like division of labour scheme. Especially during the first development phases of the platform, each partner did not have an explicitly determined role. Furthermore, at least temporarily, the number of participating actors had been excessively increased because the main partners were attempting to find more suitable specialized providers for the development of the platform's components. As a result, there was a significant knowledge accumulation in each individual component, but the effective combination of these components for the development of the integrated platform was delayed.

The effective operation of the platform in the near future demands extensive coordination and communication among the local authorities, KEDE and the supply-side actors so as to achieve the appropriate re-engineering and integration of the municipalities' internal business processes that underlie the digitized services. This means that KEDE should put additional effort into communicating the LGAF's goals and expected benefits to the politically elected decision-makers in municipalities, and manage collaboration between the interested parties in order to attain the pilot use of the system and pilot delivery of e-government services.

Finally, an extended and economically viable utilization of the LGAF platform demands the cooperation of the potential end-users and supply-side entities for the creation of a business scheme that will cover the system's operational cost. For instance, public–private partnerships in combination with outsourcing practices, constitute an efficient risk management instrument as they contribute to risk-sharing between public customers and suppliers. Generally, in complex projects, like LGAF, that produce composite products (systems) and bundles of services, partnership-based approaches are more appropriate (Schapper et al, 2006; Caldwell and Howard, 2011).

NOTES

1. This case study was partially funded by the FP7 AEGIS (Advancing Knowledge-Intensive Entrepreneurship and Innovation for Economic Growth and Social Well-being in Europe) project (contract number: 225134).
2. Interoperability is the ability of disparate and diverse organizations to interact towards mutually beneficial and agreed common goals, involving the sharing of information and knowledge between the organizations, through the business processes they support, by means of the exchange of data between their respective ICT systems (European Commission, 2010c, p. 2).
3. Specifically, the European Interoperability Strategy (European Commission, 2010b) and the European Interoperability Framework (European Commission, 2010c) are the key milestones which the Digital Agenda for Europe is based upon.
4. A sector is understood as a policy area, for example customs, police, e-health, environment, agriculture (European Commission, 2010c, p. 1).
5. The term 'life-event services' can be defined as a basket of services that are relevant in a particular point in time to a business or citizen. Such services, typically delivered in silos from individual agencies, are required to be streamlined across public (and at times private) organizations to adequately satisfy the customer (European Commission, 2013, p. 5).
6. The survey was conducted by the Laboratory of Industrial and Energy Economics of the National Technical University of Athens (LIEE/NTUA, http://www.liee-ntua.gr/en/) using a structured questionnaire.
7. http://www.kedel.org/wiki/ (accessed July 2013).
8. A standard is an agreement among independent parties about how to go about doing some task (Bloomberg and Schmelzer, 2006, p. 35). Technically, it is a framework of specifications that has been approved by a recognized organization or is generally accepted and widely used throughout by the industry (Nah, 2006, p. 1). The standards that best promote interoperability are open standards (Eric Sliman, 2002, 'Business case for open standards', http://www.openstandards.net/viewOSnet1C.jsp?showModuleName=businessCaseForOpenStandard (accessed June 2013).
9. The European Interoperability Framework (European Commission, 2010c, p. 21) defines four Interoperability levels: (i) technical: planning of technical issues involved in linking computer systems and services; (ii) semantic: precise meaning of exchanged information which is preserved and understood by all parties; (iii) organizational: coordinated processes in which different organizations achieve a previously agreed and mutually beneficial goal; and (iv) legal: aligned legislation so that exchanged data is accorded proper legal weight.
10. 'Scalability' is defined as the usability, adaptability and responsiveness of applications as requirements change and demands fluctuate (UNDP, 2007, p. 5).
11. 'Reusability' is defined as the degree to which a software module or other work product can be used in contexts other than its original, intended or main purpose (European Commission, 2010c, p. 33).

12. http://en.wikipedia.org/wiki/Open-source_software (accessed July 2013).
13. http://open-source.gbdirect.co.uk (accessed June 2013).
14. www.oss-watch.ac.uk/resources/cs-aplaws, www.oss-watch.ac.uk/resources/cs-aplaws, http://www.opensourceacademy.org.uk/solutions/casestudies/aplaws/file (accessed June 2013).
15. A notable fact reported during the interviews, which may have also intensified the need for redesign, was that the initial survey of existing technical solutions was rather narrow because it focused on the field of local government. This limited considerably the range of the proposed software packages and probably counted out more state-of-the-art technical solutions.
16. 'Bringing SOA value patterns to life: an Oracle White Paper', p. 5. See http://www.oracle.com/technologies/soa/soa-value-patterns.pdf (accessed June 2013).
17. Peter's Pensieve OASIS Symposium – Service-Oriented Architecture and e-Government Panel. See http://www.xmlbystealth.net/blog/2007/04/oasis-symposium-soa-and-egovernment.html (accessed June 2013).
18. A part of the core components of the platform would be based on existing OSS packages.
19. Apart from the work delay, another reason was that Greece had entered a period of severe economic crisis that also affected the absorption of EU funds and, consequently, the funding of several ongoing technology projects.
20. United Nations Statistics Division (UNSD) (accessed May 2013).

REFERENCES

Biebertein, N., Bose, S., Fiammente, M., Jones, K., Shah, R. (2006). *Service Oriented Architecture Compass: Business Value, Planning, and Enterprise Roadmap*. Upper Saddle, NJ: IBM Press.

Bloomberg, J., Schmelzer, R. (2006). *Service Orient or Be Doomed*. Hoboken, NJ: Wiley & Sons.

Caldwell, N., Howard, M. (2011). *Procuring Complex Performance: Studies of Innovation in Product-Service Management*. New York: Routledge.

Central Union of the Greek Municipalities (KEDE) and National Technical University of Athens (NTUA) – Laboratory of Industrial and Energy Economics (LIEE) – Research team INFOSTRAG (2011). Survey results: resources and capabilities for the functional utilization of ICTs in Greek municipalities (in Greek).

Edler, J., Georghiou, L. (2007). Public procurement and innovation – resurrecting the demand side. *Research Policy*, 36, 949–963.

Edler, J., Hommen, L., Papadokou, M., Rigby, J., Rolfstam, M., Tsipouri, L., Ruhland, S. (2005). Innovation and public procurement. Review of issues at stake. Study for the European Commission, final report.

Edquist, C., Zabala-Iturriagagoitia, J.M. (2012). Public procurement for

innovation as mission-oriented innovation policy. *Research Policy*, 41, 1757–1769.
Edquist, C., Zabala, J.M., Timmermans, B. (2010). A conceptual framework for analyzing the relations between demand and public innovative procurement and between knowledge intensive entrepreneurship and innovation. AEGIS project [EC/FP7] (Advancing Knowledge-Intensive Entrepreneurship and Innovation for Economic Growth and Social Well-being in Europe) – Deliverable 1.5.1.
European Commission (2010a). *A Digital Agenda for Europe*. COM(2010) 245 final/2, Brussels, 26.8.2010.
European Commission (2010b). *European Interoperability Strategy (EIS) for European public services*. COM (2010) 744 (Annex 1), Brussels, 16.12.2010.
European Commission (2010c). *European Interoperability Framework (EIF) for European public services*. COM (2010) 744 (Annex 2), Brussels, 16.12.2010.
European Commission (2010d). Digitizing public services in Europe: putting ambition into action – 9th Benchmark Measurement. Prepared by Capgemini, Sogeti, IDC, RAND Europe and the Danish Technological Institute for the Directorate General Information Society of the European Commission.
European Commission (2012). RIS3 National Assessment: Greece. Smart specialization as a means to foster economic renewal. Report to Directorate General for Regional Policy, Unit I3-Greece and Cyprus.
European Commission (2013). Public services online: digital by default or by detour? e-government Benchmark 2012, Final Insight Report. Prepared by Capgemini, Sogeti, IS-practice and Indigov, RAND Europe and the Danish Technological Institute for the Directorate General for Communications, Networks, Content and Technology of the European Commission.
Gallerani-Cineca, F. (2011). Build once, use many: how to recycle enterprise components to offer new online services. EUNIS International Congress, conference on Maintaining a Sustainable Future for IT in Higher Education. Dublin, Ireland, 15–17 June.
Jeong, C.H.I. (2007). *Fundamental of Development Administration*. Selangor: Scholar Press.
Malerba, F., McKelvey, M. (2010). Conceptualizing knowledge intensive entrepreneurship: concepts and models. AEGIS project [EC/FP7] – Deliverable 1.1.1.
Nah, S.H. (2006). FOSS: open standards. Asia-Pacific Development Information Programme e-Primers on Free/Open Source Software. http://www.iosn.net/open-standards/foss-open-standards-primer/foss-openstds-withcover.pdf (accessed June 2013).
Nioras, A. (2011). Mini country report/Greece. Under specific contract

for the integration of INNO Policy TrendChart with ERAWATCH (2011–2012).
Novakouski, M., Lewis, G.A. (2012). Interoperability in the e-government context. Technical note for the Research, Technology, and System Solutions Program. Software Engineering Institute.
Radosevic, S., Yoruk, E., Edquist, C., Zabala, J.M. (2010). Innovation systems and knowledge intensive entrepreneurship: analytical framework and guidelines for case study research. AEGIS project [EC/FP7] – Deliverable 2.2.1.
Schapper, P., Veiga Malta, J., Gilbert, D. (2006). An analytical framework for the management and reform of public procurement. *Journal of Public Procurement*, 6(1–2), 1–26.
SEV, IOBE, NTUA (2011). Report: evaluation and perspectives of the Greek ICT sector: results of an experts panel. Research project for the prediction of changes in regional production systems and labor markets, Hellenic Federation of Enterprises, Foundation for Economic and Industrial Research and National Technical University of Athens (in Greek).
Technopolis (2011). How public procurement can stimulate innovative services. Report to Nordic Innovation Centre.
Timmermans, B., Zabala-Iturriagagoitia, J.M. (2013). Coordinated unbundling: a way to stimulate entrepreneurship through public procurement for innovation. *Science and Public Policy* doi: 10.1093/scipol/sct023, 1–12.
United Nations Development Programme (UNDP). (2007). e-government interoperability: guide. UNDP GIF Study Group.
Valovirta, V. (2012). Towards a management framework for public procurement of innovation. Conference on Demand, Innovation and Policy: Underpinning Policy Trends with Academic Analysis. Manchester Institute of Innovation Research, 22–23 March.

9. Closing the loop: examining the case of the procurement of a sustainable innovation

Jillian Yeow, Elvira Uyarra and Sally Gee

INTRODUCTION

The potential of public procurement of innovation to address 'grand challenges', such as sustainability, is increasingly acknowledged in both policy and academic circles (Edler and Georghiou, 2007; Edquist and Zabala-Iturriagagoitia, 2012). Procurement is indeed increasingly being used to address multiple policy agendas and objectives, be these social, environmental or otherwise (Erridge, 2004; McCrudden, 2004).

This chapter explores the potential of the public sector to pursue specific sustainability goals through the procurement of innovation. This is done through an examination of a UK government initiative to collect and recover its own paper waste and produce a 'closed-loop' recycled copier paper. The model involves the shredding of confidential paper waste on-site, and the subsequent processing of this waste into recycled copier paper off-site, which is then sold back to government departments for their use. A key innovation in this process is the earmarking of paper supply for return to the (contributing) client organization. Creating a 'closed loop', the paper introduced traceability into waste disposal, ensured data security, stabilized expenditure on paper, reduced associated costs, and enabled both supplier and government client to capitalize on a burgeoning paper market. We detail how the parties involved also stimulated organizational, environmental and supply chain innovation through their procurement (purchasing) activities to achieve a more efficient, more sustainable and more cost-effective outcome.

This chapter therefore deals with the process leading to the development and co-design of the 'closed loop', and addresses the main drivers and barriers influencing this particular innovation. In so doing, we identify some of the relevant processes, mechanisms and structures for procuring sustainable innovations. Specifically, we consider the extent to which the procurement of this innovation sought to achieve multiple objectives and the way in which the public sector acted as a lead user, triggered innovation in the private sector and generated market demand for the innovation. This case was not motivated solely by sustainability objectives; yet, by taking a systemic approach, it enabled a much more deeply rooted problem to be solved, thereby satisfying multiple objectives (including sustainability) through public procurement and ultimately paving the way for greater demand in the market for recycled paper.

Through this chapter, we aim to make a contribution to the discussion of procurement as a multi-objective or multifaceted policy instrument, whilst providing a conceptual understanding of the organizational, institutional and economic drivers and barriers that have been identified in the literature as influencing the successful integration of sustainability and innovation. Theoretically, this chapter brings together two interrelated, albeit often separately considered, strands of literature dealing with innovation procurement and sustainable procurement, and presents a case for considering them in tandem.

Empirically, this case study contributes to a growing evidence base of procurement-led innovations, in a relatively underexplored, yet very relevant area for innovation and sustainability. The UK public sector is a significant consumer of copier paper and, as a result, also a major generator of paper waste. Consequently there is significant potential for the public sector to harness its purchasing power to address issues of sustainability relating to its paper consumption and waste. Given the growing impact of deforestation and increasing prices of virgin pulp to make paper, there are considerable economic and environmental benefits associated with stimulating the expansion of the recycled paper market.[1] There are several barriers to increasing the production of recycled paper, on both the demand and supply sides, including a lack of markets for collected material and a lack of incentives to encourage commercial entities to recycle. Through this case study, we highlight the relevant processes, mechanisms and structures for procuring sustainable innovations and overcoming some of these challenges.

The rest of the chapter is structured as follows. First, we

provide a brief overview of the literature on public procurement and, more notably, public procurement of innovation. In particular, we focus on the triggers and challenges to public procurement of innovation, and consider the burgeoning attention on sustainability as a procurement objective. We briefly explore the key debates that have emerged in the field of sustainable procurement, before linking the two fields of literature. We then introduce the research methodology, background to the case study, and the empirical findings. This is followed by a discussion of the results, including policy implications.

PUBLIC PROCUREMENT OF INNOVATION (PPOI)

Public procurement refers to the purchase or acquisition of goods and services by government or public sector organizations. Consequently, public procurement of innovation (PPoI) refers to the commissioning and purchasing of goods and services that are new to the purchasing organization, and which enable a novel service or more efficient or effective delivery of that service (Edler and Georghiou, 2007; Edquist and Hommen, 2000; Uyarra and Flanagan, 2010). According to HM Treasury (2012), UK public bodies spent about £238 billion on procurement of goods and services in 2010/11, accounting for 35 per cent of total public expenditure and approximately 16 per cent of the UK's gross domestic product (GDP) (Uyarra et al., 2013). Public sector procurement can have a significant effect on the dynamics of markets and competition through its large purchasing power, by creating new, forward-looking markets downstream and scientific knowledge upstream (Edler et al., 2012).

A more recent major development has been the orientation of innovation policy towards explicitly solving 'grand challenges' (Edler and Georghiou, 2007; Edquist and Zabala-Iturriagagoitia, 2012), and this can be seen in the focus and approach of the Horizon 2020 programme (European Commission, 2013). One such solution is through the role of government as the purchaser of innovation, be that responsively to an existing innovation or proactively through triggering innovative solutions by a newly specified need (Allman et al., 2011). The public sector can act as a lead user (Von Hippel, 1986) and 'use its purchasing power to influence suppliers and the products they develop and design, for the wider benefit of others in the economy and

UK environment' (NAO, 2009, p. 4). The state can also trigger a 'lead market' of, in this case, a sustainable innovation (Aho et al., 2006).

Lead markets are geographical areas that pioneer the successful adoption of an innovative design, starting a process of global diffusion (Beise, 2004). We consider that the UK public sector, when prepared to undertake high initial costs and risks associated with early adoption of an innovation, can be a lead market through its significant purchasing power (Georghiou, 2007). Innovations supported through procurement may evolve to become cheaper and more effective such that they can be rolled out to other markets.

Despite the obvious potential of procurement to achieve multiple objectives, discussions on the contribution of procurement to various policy agendas (for example, social policy, sustainability, innovation, efficiency) tend to be disconnected, with insufficient acknowledgement of the tensions and complementarities between policy objectives. For instance, Edler and Georghiou (2007) have noted that the connection between innovation and certain policy goals is 'still insufficiently examined in the literature and poorly designed and taken advantage of in policy practice' (p. 957).

SUSTAINABLE PROCUREMENT AND THE LINK WITH PPOI

Sustainable procurement is defined as 'a process whereby organisations meet their needs . . . in a way that achieves value for money on a whole life basis . . . whilst minimising damage to the environment' (Defra, 2005, p. 10). The sustainable procurement literature tends to concentrate on products as they move along the supply chain, specifically the environmental credentials of suppliers, often through life cycle analysis (LCA) or corporate social responsibility schemes. The sheer size of public sector demand in environmentally sensitive areas such as transport, energy, food and paper confers obvious potential for linking procurement and sustainability (Defra, 2006). The contribution of innovation to sustainable procurement is increasingly acknowledged; according to the National Audit Office (NAO, 2009), 'the public sector has considerable buying power, and the ability to influence supply chains to address government priorities such as sustainability both directly and ... through encouraging innovation' (p. 20). However, discussions on innovation and sustainability have often been restricted to the procurement of manufactured goods, in particular high-technology products.

Caldwell et al. (2009) argue that a manufacturing bias dominates procurement practices, when paradoxically the work of procurement professionals is increasingly characterized by purchasing a combination of products and services. They concur that a move towards integrated and more complex products and systems implies that 'the task cannot be accomplished by the serial and additive transaction mode of traditional (manufacturing) procurement' (Caldwell et al., 2009, p. 178). This is particularly pertinent in the case of sustainable procurement, and supported in the emerging literature. Walker (2010) has also taken into consideration how services can be provided by suppliers in a more sustainable way. The provision of more sustainable services usually requires the co-production of the service (dependent on both use knowledge and technology knowledge), and is realized in the users' context (Edvardsson et al., 2012; Möller et al., 2008). Post-product sale issues, such as disposal and recovery of materials, have arguably led to a shift from purchasing discrete products to purchasing a set of services. Similarly the Sustainable Procurement Task Force[2] (Defra, 2006) suggests that there must be a shift from the procurement of products to the procurement of services to deliver sustainability objectives.

Little strategic focus has been put on innovations in services and processes, particularly in non-operational and lower-value goods and services. Different strategies are needed when considering the procurement of different types of products and services, as suggested by the purchasing portfolio model developed by Kraljic (1983). Kraljic classifies products on the basis of two key dimensions: profit impact and supply risk. Accordingly, four categories of purchased items emerge: 'bottleneck', 'non-critical', 'leverage' and 'strategic'. Each of the categories would necessitate a differentiated purchasing strategy. Items in the bottleneck and strategic categories would generally be those goods and services that are of strategic importance relative to the mission and objectives of the organization. Bottleneck items require volume insurance, vendor control, security of inventories and back-up plans to reduce supply risk, while in the case of strategic items, different supplier strategies are suggested according to relative power positions: 'diversify', 'balance' or 'exploit' (Kraljic, 1983).

While Kraljic's purchasing portfolio model allows classification of an organization's spending in terms of the strategic importance of an item and the extent of supplier or market risk involved, some authors have suggested that it is also important to consider how to move around the different portfolio segments (Gelderman and Van Weele, 2005). Some critics go so far as to suggest that portfolio models have a tendency to result in strategies that are independent of each other (Coate, 1983), that they concentrate on separate products (Ritter, 2000), and do not depict the interdependencies between two or more items in a matrix.

CHALLENGES AND DRIVERS OF PPOI

The advantages of aggregation of demand include better management of information, greater leverage for contracting with suppliers, greater economies of scale and lower transaction costs (for example, by simplifying the tendering process). However, this potential (purchasing power) has not been realized due to the fragmented nature of public sector purchasing (European Commission, 2011; OFT, 2004). Even when the public sector accounts for a significant share of the total demand in a particular market, if different functions or departments are buying the same goods individually and in an uncoordinated fashion, purchasing power is ineffective. As noted by Phillips et al. (2007, p. 79), fragmentation and the co-existence of many 'different purchasing decision points . . . can result in disharmony and a reduction in . . . purchasing power'. This issue was reiterated in the Efficiency Review by Sir Philip Green (2010). Even seemingly 'unitary' parts of the public sector may not act as a coherent whole in practice (Caldwell et al., 2005), as they can comprise different decision-making and purchase points. However, large contracts do not necessarily lead to greater innovation; in fact the opposite may be true: large purchasing can lead to incumbent advantages, market distortion, narrowing of technological trajectories, even encouraging lock-in to suboptimal technologies or standards, and more conservative decision-making (Uyarra and Flanagan, 2010). Smaller lots of purchasing can, on the other hand, allow more managed risk-taking to test new innovations (Georghiou et al., 2013).

Some authors have identified procurer competence as a concern when procuring innovative solutions. They point to the possible discrepancy between the capabilities held by procurers and the skills required for procuring innovative solutions (Yeow and Edler, 2012). As noted by Rothwell and Zegveld (1981), whereas relatively little in-house competence is needed when procuring off-the-shelf goods for the lowest possible price, greater competence is required to encourage suppliers to innovate. Changes in the procurement function towards a more strategic orientation, and a more demanding environment for procurement, have led commentators to critically examine the skill and competency requirements of procurement professionals (Tassabehji and Moorhouse, 2008). Cousins et al. (2006b) found that purchasers with high skill levels and knowledge have a significant impact on financial performance and operational efficiency in terms of quality improvement, design and reduction of lead times. They differentiate strategic, celebrity, undeveloped and capable purchasing, according to their performance in the following: performance against strategic planning, purchasing skills, purchasing status and internal integration. The Sustainable Procurement Task Force noted that many parts of the public sector lacked professional procurement expertise (Defra, 2006, p. 47). In particular, there was a lack of understanding about sustainability and its relationship to procurement; they commented that this was partly due to the fact that environmental specialists rather than procurement experts deliver sustainable procurement training.

Numerous studies have also highlighted the position of procurement in the internal hierarchy of the organization and its status relative to other corporate functions. Despite procurement being increasingly seen as strategic in both public and private organizations, a review by Zheng et al. (2007) suggests an uneven picture in relation to procurers' influence over corporate-level strategic decisions or make-or-buy decisions, and in managing relations with suppliers. The status of the procurement or purchasing function tends to be lower than in other functional areas, particularly in the public sector (Uyarra, 2010). This relatively low influence is aggravated by a general lack of commitment and ownership of procurement strategies by senior management and political leaders (Defra, 2006; Morgan, 2008; Walker and Brammer, 2009). Defra (2006) found that there was a lack of clear direction from top management to make delivering sustainable development objectives through procurement a priority. Related to the importance of leadership, the role of champions has been identified in securing the success of certain innovations, such as the introduction of digital signal process hearing aids into the NHS as reported by Phillips et al. (2007). A champion is a 'charismatic individual(s) who throws his or her weight behind an innovation, thus overcoming indifference or resistance that the new idea may provoke in the organisation' (Rogers, 1995, p. 414). Champions are typically powerful individuals high in the management of an organization.

The availability and quality of procurement data in the public sector have also been identified as challenges to procuring sustainable solutions. A National Audit Office (NAO, 2009) report on the environmental impacts of public procurement referred to the lack of data availability as a significant barrier to managing demand by procurement teams. HM Treasury (2009) stressed that, in many parts of the public sector, information on what is spent is of insufficient quality to support decision-making and ensure progress against policy agendas. Spending cannot be effectively managed if it cannot be articulated effectively in the first place (HM Treasury, 2009, p. 20). Diversity in accounting structures, uneven data availability, a lack of widely accepted data standards and insufficient use of technology are among the key barriers preventing good management information on procurement.

A good working relationship between buying and supplying organizations has also been highlighted as important in order to reduce uncertainty and encourage innovative responses from suppliers. Proximity, trust and reciprocity are all features of learning and related to innovation (Edler and Gee, 2013). Partnerships have the potential to build social capital by developing long-term relationships with private sector suppliers. Erridge and Nondi (1994) argued that interaction and exchange lead to developing trust and shared norms that reduce opportunism, the need for costly monitoring and general transactions costs associated with exchange in instances where there is information asymmetry. Zheng et al. (2008) analysed the interplay between relational and contractual governance in relation to two public finance initiative case studies, and concluded that relational and contractual mechanisms are complementary. They found that contractual governance capability was insufficient for effective exchange and needed to be complemented by proactive relational governance based upon interpersonal trust. Similarly, Cousins et al. (2006a) found that informal socialization processes were important in creating relational capital whereas formal socialization processes were less effective.

Nonetheless, such buyer–supplier relationships can be hindered by procurement practices, as suggested by Erridge and Greer (2002), who identified a number of barriers preventing a partnership approach to public procurement. They note that 'regulations and rules to ensure financial probity and competitive tendering have restricted the development of closer supply relations and social capital by setting out rigid bureaucratic procedures, and creating a public sector culture which is risk averse and resistant to change' (p. 519). They suggest that there is a situation of imbalance between transparency, value for money and relationship development, driven by rigid rules and bureaucratic processes, low levels of procurement expertise, a lack of interdepartmental collaboration and little involvement of senior departmental managers. According to Erridge and Nondi (1994), procurement practices that prevent adequate public–private partnering include 'rigid application of tendering procedures for low-value items regardless of non-costs; too many suppliers; short-term contracts and the absence of cooperation from suppliers' (p. 178). This focus on transparency when combined with a risk-orientated culture manifests in high levels of contractual procedures leading to reduced flexibility, trust, experimentation and ultimately innovation.

MAKING THE CASE FOR PROCUREMENT OF A SUSTAINABLE INNOVATION: THE CASE OF CLOSED-LOOP PAPER

Context and Methodology

HM Revenue and Customs (HMRC) is one of the largest central government departments in the UK public sector. Commercial spending in HMRC is approximately £2 billion, of which £600 million is spent on the procurement of third-party goods and services (HMRC, 2010b). In November 2007, HMRC hit the news headlines when it was reported that two discs containing the personal details of 25 million people had been lost in transit to the National Audit Office. This incident was considered to be catastrophic, triggering both an independent and a major internal review. These reviews exposed serious issues at HMRC, including: low priority accorded to information security, a lack of management accountability, an overly complex organizational structure, and highly fragmented and unaccountable waste management contracts throughout its estates. The loss of personal data forced issues of confidentiality and waste management up the political and managerial agenda, and created a sense of urgency at the departmental level. The incident was a catalyst for innovation, and resulted in the invention and deployment of a 'closed-loop paper' model, which forms the basis of this case study. The case study that follows charts the antecedents, processes and outcomes of the closed-loop paper initiative. The distinctive circumstances and characteristics of this case led to the development of a solution that was unique to this setting.

A case study approach was chosen to allow a 'richly detailed portrait of a particular phenomenon' (Hakim, 2000, p. 59) and focus on the dynamics in a single setting within a real-life context (Yin, 1994, 2003). While the UK public sector is very large, it is also very varied. Although multiple cases provide additional evidence to support the resulting analysis, single cases allow for in-depth investigation and rich description of a phenomenon (Walsham, 1995). A single-case focus enables highly detailed examination of process, and is appropriate when a case is extreme, unique or revelatory (Yin, 1994, pp. 38–40).

Multiple sources of evidence have been used to construct the case study. Data is drawn from in-depth interviews with public and private stakeholders participating in the process as well as observations, site visits and secondary data sources. Semi-structured interviews were conducted with all key individuals involved in the case in an iterative process. This includes those at senior management levels in the procurement function of the buyer as well as the supplier organizations. The interviews were supplemented with site visits by the researchers to both procuring and supplier organizations based in the UK, as well as the paper mill, located in Germany. An analytical line of enquiry was developed prior to interviewing in order to reduce bias and reflexivity, and background research conducted to inform the interview process. Questions asked explored general issues of procurement organization and processes within HMRC, challenges faced in public sector procurement (by both procuring and supplier organizations), the processes undertaken in the transformation of the procurement of this particular innovation, and the key factors that were integral in enabling this transformation. Data was collected between June 2011 and January 2012. The case study also draws on a variety of secondary data sources including market analysis, government reports, academic and trade journal articles and media coverage. The data has been triangulated in order to reduce some problems associated with case study interviews, for example bias and selective memory. The case is presented as a narrative.

Case Background: Traditional Approach to Waste Management and Paper Procurement

One of the ways in which the UK public sector has attempted to capitalize on its purchasing power is to use framework agreements and consortiums when purchasing low-risk items such as copier paper to ensure that prices remain competitive. However, that is only one side of the story; the paper that is used also needs to be disposed of. A depiction of a paper supply chain often begins from the forest (or paper mill) and ends at the production of paper products (or consumption by the end-user) (see Figure 9.1). There is often no consideration of what happens to the paper after it is used. Instead, paper consumption then forms the beginning of a different supply chain, that of (paper) waste disposal and recovery (Figure 9.2). Consequently, the purchase and use of paper products and then its subsequent disposal and recovery are often considered, and therefore procured, separately.[3] We argue that there is potential to consider the purchase and disposal of paper in a more holistic manner, with the end-user organization at the centre, acting as both a buyer of paper products as well as a supplier of paper waste, which could potentially be turned into paper products (that they also buy). Considering that the majority of HMRC's activity involves VAT returns, and therefore that the department is heavily reliant on paper-based communication,[4] it is thus not only a significant consumer of copier paper, but also a significant producer of paper waste, and has the potential to act as a lead user in this innovation process.

Based on Kraljic's (1983) purchasing portfolio model, stationery items like copier paper would typically be considered

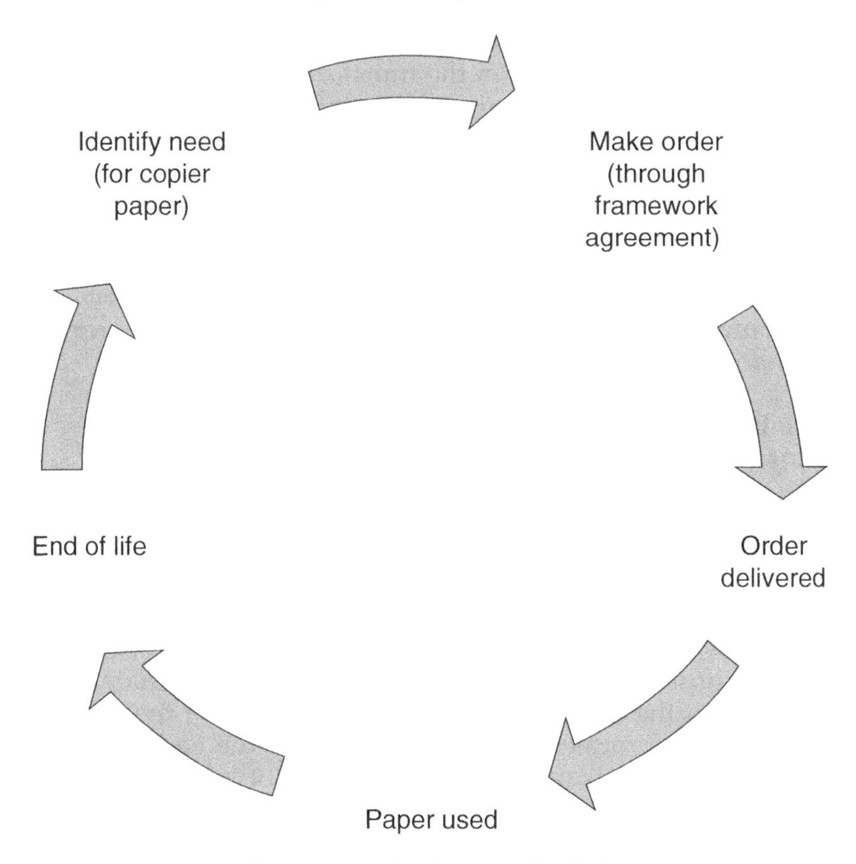

Figure 9.1 Typical paper purchasing supply chain

as a non-critical item, low in both profit impact and supply risk, important to an organization but only in terms of the smooth operational running of the organization. Based on the low value attributed to copier paper, the strategy would typically be to reduce high transaction costs (due to high order frequency) through category management and/or e-procurement mechanisms. Office supplies are considered a standard commodity, best consolidated and purchased collaboratively, implying standardizing specifications, aggregating demand and utilizing economies of scale (OGC, 2007). The UK government has taken a collaborative approach towards the procurement of office supplies, whereby several departments collaborate to jointly purchase items like stationery, copier paper and information

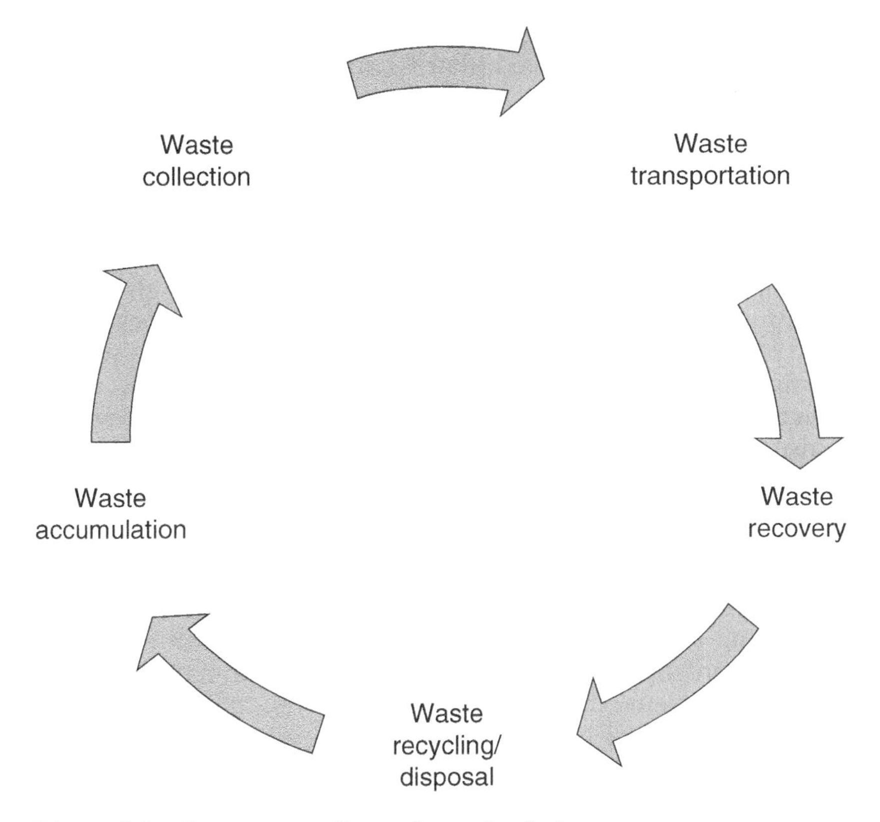

Figure 9.2 Paper waste disposal supply chain

technology (IT) consumables. In this way, procurers can achieve economies of scale through aggregated demand and negotiate better deals with suppliers. This suggests a strategic approach to increasing the profit impact of these commodities through exploitation of the purchasing power of the public sector. Using Kraljic's model to conduct an analysis of the purchasing strategies of an organization can be useful. However, this logic can break down, for example when it reinforces discrete purchasing rather than an integrated, more holistic, provision of services that can achieve better value for money. What constitutes non-critical or strategic items is relative to the mission and objectives of the incumbent organization. In the case of HMRC, paper was a leverage item due to its potential for high profit impact as a waste material (as a result of high-volume usage). Therefore its purchasing strategy could be considered differently to how paper is typically perceived (that is, non-critical) to take advantage of its value.

In contrast to paper procurement, the procurement of waste management services is often considered to be a strategic item. According to Kraljic's model, items in this category have high profitability, but also have a high supply risk. Such items require a collaborative strategy between both buyer and seller. Waste management services often require a high capital investment and long-term arrangements; in the UK these tend to be private finance initiative (PFI) contracts. PFIs are a way of creating public–private partnerships by funding public infrastructure (or services) projects with private capital. However, it is also important to note that waste management services contracts tend to be disjointed: often there will be separate contracts for waste collection, transportation, recovery and disposal; furthermore, PFI waste management arrangements are more prominent in municipal waste discussions.[5] Additionally, waste management arrangements are often specific to a particular site depending on usage and demand. Therefore it is not unusual for an organization with several different sites to have different (waste management) contracts for each site, and these might also be outsourced to the facilities management services. This was the case in HMRC, which is further elaborated in the next section.

In HMRC, the procurement of copier paper and the procurement of waste management services were dealt with separately. The procurement of paper was fully bound in its own cycle, that is, a need for copier paper arises, paper is purchased from a supplier through a framework agreement (or several framework agreements), copier paper is received and used (and then disposed of), as depicted in Figure 9.1. Waste management services were similarly managed under a separate and clearly bound arrangement (albeit to varying degrees depending on the contracts), as depicted in Figure 9.2.

It can clearly be seen from Figures 9.1 and 9.2 that it is possible to link these discrete products and services together in a way that would achieve efficiency and value for money, by streamlining the processes and purchasing activities since they have a common element: they are all related to the material paper. This is the crux of the closed-loop concept. The next section describes how a data loss incident, along with an undercurrent organizational changes occurring around the same time, allowed such changes to be enacted in HMRC.

The Closed-Loop Procurement Process

The data loss incident that occurred in HMRC in 2007 identified several structural weaknesses and obstacles in HMRC. One of the recommendations from the Poynter Review stated:

> HMRC has insufficient knowledge and oversight over its third parties' compliance with information security requirements. It should urgently address this through a programme of assurance via Internal Audit, or if they do not have the capacity via an independent third party. This should start with third parties who handle post, confidential waste, off-site storage and who provide security services. (R35)

When HMRC undertook its internal audit as recommended, it became apparent that the large number of estates that HMRC owned, and the way they were managed, posed a problem. At the time the organization had over 400 offices throughout the UK, 80 per cent of which were managed through at least five different PFI arrangements and a range of different outsourced arrangements, resulting in limited traceability of (confidential) waste once it left the organization's premises.[6] Furthermore, the budget for waste collection was not held centrally by the Commercial Department. Even if HMRC could account for how and when confidential waste left HMRC, it had very little or almost no control over what its waste disposal supplier did once it fulfilled that waste disposal contract. There was no guarantee that its disposed confidential waste, albeit disposed in a safe manner, did not end up as landfill.[7]

Against this backdrop, the aim for HMRC was to find a waste management solution that could serve all HMRC estates while avoiding a large number of individual contracts, and to facilitate tracking and tracing of their confidential waste. Recycling was also becoming a key component of HMRC's sustainable development agenda. Around the same time, the value attributed to paper waste also changed,[8] leading HMRC to recognize waste paper as a valuable commodity, and that the value of this resource was being realized by the waste suppliers, rather than the department. Furthermore, since paper also represented the largest component of HMRC's waste, reducing paper waste could potentially reduce the consumption of other stationery resources, such as toner cartridges. As such, the Commercial Department started to explore solutions to all these problems in a more efficient, systematic and joined-up way. These issues helped to frame the new procurement process for the 'closed-loop' paper solution.

The idea of transforming HMRC waste paper into 100 per cent recycled copier paper (that is, the concept known as 'closed loop') first emerged back in 2002. The concept emerged during discussions between HMRC and one of its suppliers, Banner, which supplied copier paper and other office supplies to HMRC (and several other government departments through framework agreements). Whilst most procurement relationships tended to be arm's-length in nature and embodied adversarial type interactions to negotiate the lowest possible unit price, meetings between HMRC and Banner were not limited to contract renewal and price renegotiation. Both sides often talked about how the working relationship could develop, and it was during those discussions that the closed-loop concept came about. However, at the time, the concept was very much 'blue-sky thinking' and the idea was 'parked' as several challenges were faced, including getting management buy-in, having access to a paper mill to produce it and generating enough waste paper to make the whole process viable. In particular, the latter was a major issue due to the existence of various waste disposal and collection contracts.

The existing relationship between HMRC and Banner enabled discussions about innovative ways to tackle some of the outstanding issues that existed, as well as some of the problems that emerged in 2007. Due to the extraordinary circumstances surrounding the data loss incident and the urgency of the matter, HMRC gave notice to its suppliers and contracted Banner to collect all confidential paper waste in the interim to gain some control over the situation. Banner were also tasked with surveying the waste management arrangements across the entire HMRC estate,[9] including collecting data on the volume of waste generated, and the frequency and cost of collection. Within the interim arrangement, HMRC and Banner developed, in partnership, an innovative, closed-loop solution to the confidentiality, paper waste disposal, recycling and paper supply problem that they had previously mused about but could now put into action.

Banner used the data they were collecting, in combination with an IT logistics system to organize the waste collection, and entered into a joint venture with a shredding company. Since it already worked with a paper mill based in Germany, from which it sourced its paper supply, Banner developed that relationship to become a supplier of paper pulp to the mill as well. The 'closed-loop' solution eventually developed was an integrated system of confidential paper waste collection, disposal, recovery and subsequent supply of 100 per cent recycled paper made entirely from government waste at a much lower price, since the raw material was provided by the customer. The paper flow became a 'closed loop' with zero waste (see Figures 9.3 and 9.4). The confidential paper waste was shredded on-site, securely baled and tagged with management information off-site and transported to a recycled paper mill from where the trucks would return with closed-loop paper produced from the previous batch. Tagging enabled the paper to be traced throughout the process, ensuring the department's waste became the department's paper supply. The recycled paper mill in turn operated a closed-loop system to a high environmental specification.[10] The closed loop is secured by the mill using a dedicated run and stopping regular lines while the dedicated production is running. The pilot closed-loop concept was successfully trialled in May 2010, and in July 2011 a contract was awarded in which the closed-loop concept was an integral element.

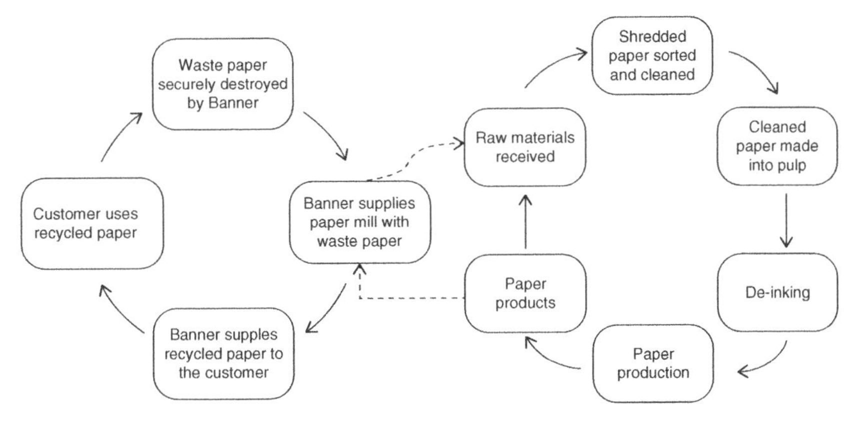

Figure 9.3 *Closed-loop paper cycle delivered by Banner*

Figure 9.4 *Closed-loop paper production at paper mill*

The closed-loop process met a number of objectives, not least resolving the confidentiality problem. Traceability and accountability of confidential paper waste were achieved; the chance of fraud was reduced and the security of information was increased as a result of confidential waste remaining in the (government or public sector) system. This objective reflected the data loss crisis, and was a primary driver for the procurement, enabling the implementation of a relatively radical innovation, involving organizational change. However, other procurement objectives were important, creating independent but mutually reinforcing drivers that informed the procurement process. Value for money was also achieved; the supplier was able to offer recycled paper at a lower (and consistent) price, as the client organization provided the raw material and the price of the recycled paper offered was not affected by price fluctuations in international recycling markets. Additionally, a competitive price for waste disposal was secured, again ensuring a financially viable innovative option.

From a sustainability perspective, the new process was more sustainable than the previous system and the environmental impact was also reduced.[11] In 2013, all central departments were mandated to use copier paper with 100 per cent recycled content; the concept of using public sector waste to produce a bespoke government-branded copier paper far exceeds this mandate (HMRC, 2010a). Overall, the process not only reduced their environmental impact, but also achieved savings of approximately £65 000 per annum (HMRC, 2010b).

At the organizational level, HMRC was able to review its processes and streamline their restricted paper waste disposal arrangements, thereby reducing unnecessary costs. Crucially, it allowed them to become more accountable through being able to track and trace the waste after it left the organization to ensure it was not compromised. These functions met a number of internal objectives, creating alignment across procurement and management objectives.

HMRC as a lead user has been working with other central government departments to help the latter make that change. The closed loop case has the potential to become a case on lead markets, that is, geographical areas that pioneer the adoption innovations, stimulating a process of global diffusion (Beise, 2004). HMRC co-generated an innovation with a supplier, bearing the initial risks and high costs associated with experimentation and innovation. The resulting innovation has subsequently been adopted by other central government departments, cumulatively supporting the emergence of a procurement-based lead market.

DISCUSSION

Due to heightened security considerations after the data loss incident in 2007, the procurement of confidential paper waste disposal services and copier paper for internal use became highly strategic (based on Kraljic's model). Indeed, the highly publicized incidents regarding the loss of confidential and personal data proved to be a significant turning point, with data security becoming a critical issue for HMRC. Internally, it was recognized that this incident provided an opportunity to turn things around in the organization and implement a more radical solution.

Due to the nature of HMRC's work, paper was a key category of spending for the organization and therefore a leverage item, which is typically characterized by its high profit impact and low supply risk, rather than merely being relegated to a low profit impact, low supply risk characteristic of non-critical items. Similarly, while waste management is typically a strategic item characterized by a high supply risk and high profit impact, in HMRC it represented a bottleneck item: high supply risk and low profit impact, held to ransom and 'locked in' to incumbent suppliers through PFI contracts that were not value for money, nor provided full accountability. As Seadon (2010) comments, '[conventionally], waste is treated as irrelevant . . . only to be managed when the pressure to handle the problem is greater than the convenience of disposal' (p. 1639). Furthermore, solutions offered are often used to satisfy short-term goals (for example, efficiency, value for money) rather than for sustainability. However, 'when waste is seen as part of a production system, the relationship of waste to other parts of the system is revealed and thus the potential for greater sustainability of the operation increases' (Seadon, 2010, p. 1641). While national (and regional) paper recycling schemes exist to divert waste from landfills, these appear to be mainly for municipal waste, and commercial entities like HMRC (despite being a public sector organization) have to find their own (commercial) ways of disposing of waste and contributing to recycling rates and landfill reduction. The 2007 data loss incident was the catalyst to manage the problem of (confidential) waste collection and disposal, whilst being able to deliver on the sustainability agenda, and to do so through innovation and, more specifically, procurement of innovation. This study shows how HMRC went one step further by not only identifying copier paper as a leverage item (instead of just a routine, non-critical item), but also changing its composition, that is, 100 per cent recycled paper made from government waste, and thus becoming a lead user in an eco-innovation and becoming both the buyer and supplier of its own paper. Here, we discuss some of the key issues and elements that were crucial in leading to the procurement of a sustainable innovation in relation to the challenges and drivers to PPoI mentioned earlier in the chapter.

In this case, by formulating a new need rather than by prescribing a particular solution, a new innovation cycle was set in motion. Innovation was induced in both buyer and supplier organizations and a new business model was created. For HMRC, working with the supplier on the pilot allowed them to review their processes and streamline their restricted paper waste arrangements; collections were only received when bins were full, which reduced their environmental impact whilst achieving savings of approximately £65 000 per annum (HMRC, 2010b). Similarly, procurement stimulated supplier innovation. In order to take up the (pilot) contract, the supplier made substantial capital investments in new IT systems, bespoke vehicles and shredding technologies. The supplier also entered into a joint venture with a shredding company, and developed a strategic partnership with a paper mill to carry out the closed-loop production run.

As previously cited, a good working relationship between buyer and supplier is often crucial to reduce uncertainty, increase trust and encourage innovation. In this case, the nature and extent of the relationship between the buyer and supplier proved crucial for enabling innovation. The longer-term relationship between the public client and the supplier enabled more open discussions. Banner had supplied paper to HMRC since 2002 and renewal meetings were not simply about renegotiating the contract price. Both parties were able to discuss how the partnership could develop. This type of relationship was found to be more conducive to identifying and developing mutually beneficial opportunities. The relationship was based on trust and transparency, and was exemplified by a pricing model based on a percentage profit for the supplier. This format differed from the pricing models of other suppliers. By working on costs and potential profits, an innovative solution economically attractive to both the client and the supplier was co-generated. This collaborative relationship through market-based interactions went beyond the transactional.

As discussed earlier in this chapter, project champions are often crucial for securing the success of certain innovations (Phillips et al., 2007). Typically, these would be individuals in positions of power within the organization. In HMRC, the main project champions were the commercial director and procurement manager. The commercial director understood the importance of procurement, and having had experience in purchasing in the private sector, recognized the disparity between private and public sector procurement and was able to bridge the gap in practices. He gave HMRC's procurement manager the flexibility, space and time needed to develop a solution. The procurement manager then engaged with suppliers, and particularly with one trusted supplier, to develop solutions

Here, we can see that procurers have an important role to play, when they are well positioned within some organizations, in influencing decisions and delivering on sustainability commitments. In this case, the procurement manager had been in this role for a long time, and had previously chaired the OGC-Buying Solutions collaborative board for office supplies. She had the necessary skills, knowledge and capability to negotiate the complicated procurement landscape. The procurement manager engaged individually with suppliers (both existing and potential) to generate ideas through 'blue-sky thinking' about how to stimulate the market, and she raised the possibility of the closed-loop concept.

Importantly, the politically sensitive nature of data security ensured senior management buy-in. This critical problem provided a space for innovation within the normal procurement process; close collaboration and coordination was possible. In this case, the creation of a temporary space in which buyer and supplier could innovate did not detract from the optimal use of resources, but utilized slack to promote experimentation and the pursuit of a potentially risky project. In normal conditions, the focus on efficiency in the public sector arguably reduces this slack, ultimately eliminating the potential of experimentation necessary for normal economic growth, and paradoxically resulting in inefficiency (Potts, 2009).

Finally, in order to address the fragmentation of demand and to control off-contract spending, HMRC moved to centralized budgets.[12] A shift towards a centralized procurement model was complemented by substantial investments in management information systems (MIS) to facilitate spending analysis and contract management.

CONCLUSION

This chapter has explored the procurement of a sustainable innovation that was initially brought about in response to a data loss incident, but which also satisfied multiple objectives and induced several other spillover effects. The emergence of a critical problem provided the context and impetus for change, and offered an opportunity not just to do things better but also to do things differently, which the organization achieved through the procurement of a sustainable innovation.

The conventional approach to waste management delinks waste generation, collection and disposal systems as independent operations despite the close linkages between them (Seadon, 2010). From an environmental or sustainability perspective, such a reductionist approach is not helpful. It also assumes that the different types of waste (and their management) are not related. In the 'closed-loop' case, the original objective was not related to innovation or sustainability, but by viewing the problem from a systemic perspective and considering the viability of an integrated solution instead of focusing solely on addressing the individual problems that emerged, ultimately innovation was triggered, which achieved results over and above the original criteria.

This was crucially enabled by the right mix of skills and capabilities within the organization, in this case within the procurement team, and also by according such personnel the trust and space within which to develop the concept and operationalize it. Other enabling factors in this process are, for example, the influence of a product champion and a good working relationship between the procurer and the supplier, including a mutually beneficial rewards model.

The case provides several implications for policy. It suggests that the procurement of innovation can be best enabled by a systemic and service-oriented approach. The closed-loop case has shown that innovation tends to be associated with procurement that is linked to activities with true leverage or critical to the pursuit of the organization's strategic objectives, and has the potential to send a clear and consistent signal to the market once a need and its outcome have been demonstrated. This case represents an exemplar of demand-side innovation measures, which can enable enterprises to gain a better return on their innovation efforts and, through public demand, spur the market to self-organize and further anticipate innovation demands.

Nonetheless, we recognize that the use of a single case study is not without its limitations. Hartley (2004, p. 326) points out that 'the challenge (in single case studies) is to disentangle what is unique to that organisation from what is common to other organisations', and the findings are not generalizable. However, we believe that this case represents a rich and unique case from which lessons may be learnt and applied to other parts of the public sector.

NOTES

1. Recycling paper not only reduces landfill waste; its other benefits (in addition to virgin paper production and incinerating it for energy) include reduced pollution and energy use.
2. The Sustainable Procurement Task Force, jointly funded by the Department for the Environment, Food and Rural Affairs and HM Treasury, was set up in 2005 to devise a National Action Plan to deliver the objective of making the UK a leader in the European Union (EU) in sustainable procurement by 2009.
3. Additionally, waste management contracts are often unique to the estate and negotiated individually, resulting in an organization with several locations operating multiple waste management contracts.
4. In 2009/10, HMRC's spending on printing, postage, stationery and office supplies was £104.9 million.
5. Whilst a large amount of literature exists on municipal and household waste, statistics relating to commercial waste produced by public sector organizations in England are scarce, except statements stating that government estates aim to reduce their waste output. According to Defra, 47.9 million tonnes of commercial and industrial waste were generated in 2009 (Defra, 2010); no specific data was available for waste generated by the public sector.

6. It was estimated that there were at least 43 different arrangements in place for restricted and confidential waste disposal throughout the organization.
7. Similarly, suppliers (of waste collection and disposal services) may offer to recycle the waste paper, but that may merely mean selling it off to a recycled paper mill on HMRC's behalf, and thus diverting profits from the sale of HMRC paper waste directly to a mill.
8. The typical price of recovered paper was around £40–£60 per tonne in 2007.
9. As part of a change programme running concurrently (in 2006 and 2008), HMRC consolidated its estates down to 235 offices, and relocated all procurement staff to a single office (in Salford, Greater Manchester). The decision to centralize procurement was perceived by the commercial director as important to the effective leverage of the function.
10. The mill, based in Germany (there are currently no UK paper mills able to perform this function), uses 100 per cent less wood pulp to make paper (compared to virgin paper); up to 83 per cent less water consumption; up to 72 per cent reduction in energy use; up to 46 per cent reduction in carbon dioxide (CO_2) emissions; and an integrated (combined heat and power) plant that has 50 per cent reduced CO_2 emissions and 87 per cent increased thermal efficiency.
11. There is potential, if rolled out, or diffused, for this process to shape the evolution of the recycled paper market in the UK.
12. HMRC's procurement capability review (conducted in 2008) reports that 'maverick' spending is less than 2 per cent; in comparison, the US Internal Revenue Service reports maverick spending of less than 0.5 per cent.

REFERENCES

Aho, E., Cornu, J., Georghiou, L., Subira, A. (2006). Creating an innovative Europe. Report of the Independent Expert Group on R&D and Innovation appointed following the Hampton Court Summit, http://ec.europa.eu/invest-in-research/action/2006_ahogroup_en.htm.

Allman, K., Edler, J., Georghiou, L., Jones, B., Miles, I., Omidvar, O., Ramlogan, R., Rigby, J. (2011). Measuring wider framework conditions for successful innovation: a system's review of UK and international innovation data. London: NESTA.

Beise, M. (2004). Lead markets: country-specific drivers of the global diffusion of innovations. *Research Policy*, 33(6–7), 997–1018.

Caldwell, N.D., Roehrich, J.K., Davies, A.C. (2009). Procuring complex performance in construction: London Heathrow Terminal 5 and a Private Finance Initiative hospital. *Journal of Purchasing and Supply Management*, 15(3), 178–186.

Caldwell, N., Walker, H., Harland, C., Knight, L., Zheng, J., Wakeley, T. (2005). Promoting competitive markets: the role of public procurement. *Journal of Purchasing and Supply Management*, 11(5–6), 242–251.

Coate, M.B. (1983). Pitfalls in portfolio planning. *Long Range Planning*, 16(3), 47–56.
Cousins, P.D., Handfield, R.B., Lawson, B., Petersen, K.J. (2006a). Creating supply chain relational capital: the impact of formal and informal socialization processes. *Journal of Operations Management*, 24(6), 851–863.
Cousins, P.D., Lawson, B., Squire, B. (2006b). An empirical taxonomy of purchasing functions. *International Journal of Operations and Production Management*, 26(7), 775–794.
Defra (2005). Securing the future: delivering the UK sustainable development strategy.
Defra (2006). Procuring the future – Sustainable Procurement National Action Plan: recommendations from the Sustainable Procurement Task Force.
Defra (2010). Impact assessment of proposal to revise quick wins specification for paper products.
Edler, J., Gee, S. (2013). Public procurement and the co-production of process innovation. Unpublished manuscript. Manchester.
Edler, J., Georghiou, L. (2007). Public procurement and innovation – resurrecting the demand side. *Research Policy*, 36(7), 949–963.
Edler, J., Georghiou, L., Blind, K., Uyarra, E. (2012). Evaluating the demand side: new challenges for evaluation. *Research Evaluation*, 21(1), 33–47.
Edquist, C., Hommen, L. (2000). Public technology procurement and innovation theory. In Edquist, C., Hommen, L., Tsipouri L. (eds), *Public Technology Procurement and Innovation*. Boston, MA: Kluwer Academic Publishers, pp. 5–70.
Edquist, C., Zabala-Iturriagagoitia, J.M. (2012). Public procurement for innovation as mission-oriented innovation policy. *Research Policy*, 41(10), 1757–1769.
Edvardsson, B., Kristensson, P., Magnusson, P., Sundström, E. (2012). Customer integration in service development and innovation – methods and a new framework. *Technovation*, 32(7–8), 419–429.
Erridge, A. (2004). UK public procurement policy and the delivery of public value. In Thai, K. (ed.), *Challenges in Public Procurement: An International Perspective*. Florida: PrAcademics Press, pp. 335–352.
Erridge, A., Greer, J. (2002). Partnerships and public procurement: building social capital through supply relations. *Public Adminstration*, 80(3), 503–522.
Erridge, A., Nondi, R. (1994). Public procurement, competition and partnership. *European Journal of Purchasing and Supply Management*, 1, 169–179.
European Commission (2011). Green paper on the modernization of EU public procurement policy: Towards a more efficient European

Procurement Market (Vol. COM(2011) 15 final). Brussels: European Commission.
European Commission (2013). Tackling Societal Challenges Retrieved 30th October 2013, from http://ec.europa.eu/research/horizon2020/index_en.cfm?pg=better-society.
Gelderman, C.J., Van Weele, A.J. (2005). Purchasing portfolio models: a critique and update. *Journal of Supply Chain Management*, 41(3), 19–28.
Georghiou, L. (2007). Demanding innovation: lead markets, public procurement and innovation. *NESTA Provocation 02*. London: NESTA.
Georghiou, L., Edler, J., Uyarra, E., Yeow, J. (2013). Policy instruments for public procurement of innovation: choice, design and assessment. *Technological Forecasting and Social Change*, 86, 1–12.
Green, P. (2010). Efficiency review – key findings and recommendations. http://www.cabinetoffice.gov.uk/sites/default/files/resources/sirphilipgreenreview.pdf.
Hakim, C. (2000). *Research Design: Successful Designs for Social and Economic Research*, 2nd edn. London: Routledge.
Hartley, J. (2004). Case study research. In C. Cassell and G. Symon (eds), *Essential Guide to Qualitative Methods in Organizational Research*. London: Sage, pp. 323–334.
HM Treasury (2009). Operational Efficiency Programme: final report.
HM Treasury (2012). Public expenditure statistical analyses 2012. http://www.hm-treasury.gov.uk/d/pesa_complete_2012.pdf.
HMRC (2010a). *HMRC Sustainable Procurement Strategy*. http://www.hmrc.gov.uk/about/corporate-responsibility/sustainable-proc-strategy.htm.
HMRC (2010b). *Innovation in Procurement Plan (IPP)*. http://www.hmrc.gov.uk/about/procurement.htm.
Kraljic, P. (1983). Purchasing must become supply management. *Harvard Business Review*, 61(5), 109–117.
McCrudden, C. (2004). Using public procurement to achieve social outcomes. *Natural Resources Forum*, 28(4), 257–267.
Möller, K., Rajala, R., Westerlund, M. (2008). Service innovation myopia? A new recipe for client provider value creation. *California Management Review*, 50(3), 31–48.
Morgan, K. (2008). Greening the realm: Sustainable food chains and the public plate. *Regional Studies*, 42(9), 1237–1250.
NAO (2009). Addressing the environmental impacts of government procurement.
OFT (2004). Assessing the impact of public sector procurement on competition.
OGC (2007). Saving money on office stationery. http://www.ogc.gov.uk/documents/CP0150SavingMoneyOnOfficeStationery.pdf.

Phillips, W., Knight, L., Caldwell, N., Warrington, J. (2007). Policy through procurement – the introduction of digital signal process (DSP) hearing aids into the English NHS. *Health Policy*, 80(1), 77–85.
Potts, J. (2009). The innovation deficit in public services: the curious problem of too much efficiency and not enough waste and failure. *Innovation: Management, Policy and Practice*, 11(1), 34–43.
Ritter, T. (2000). A framework for analyzing interconnectedness of relationships. *Industrial Marketing Management*, 29(4), 317–326.
Rogers, E. (1995). *Diffusion of Innovations*. New York: Free Press.
Rothwell, R., Zegveld, W. (1981). *Industrial Innovation and Public Policy: Preparing for the 1980s and the 1990s*. Westport, CT: Greenwood Press.
Seadon, J.K. (2010). Sustainable waste management systems. *Journal of Cleaner Production*, 18(16–17), 1639–1651.
Tassabehji, R., Moorhouse, A. (2008). The changing role of procurement: developing professional effectiveness. *Journal of Purchasing and Supply Management*, 14(1), 55–68.
Uyarra, E. (2010). Opportunities for innovation through local government procurement. London: NESTA.
Uyarra, E., Edler, J., Gee, S., Georghiou, L., Yeow, J. (2013). UK public procurement of innovation: the UK case. In Lember, V., Kattel, R., Kalvet, T. (eds), *Public Procurement, Innovation and Policy: International Perspectives*. London: Springer-Verlag, pp. 233–258.
Uyarra, E., Flanagan, K. (2010). Understanding the innovation impacts of public procurement. *European Planning Studies*, 18(1), 123–143.
Von Hippel, E. (1986). Lead users: a source of novel product concepts. *Management Science*, 32(7), 791–805.
Walker, H. (2010). Successful business and procurement: What lessons for sustainable public procurement can be drawn from successful businesses? http://www.sd-research.org.uk/wp-content/uploads/high-res-procurement-full-review-8dec2010-formattedx.pdf.
Walker, H., Brammer, S. (2009). Sustainable procurement in the United Kingdom public sector. *Supply Chain Management: An International Journal*, 14(2), 128–137.
Walsham, G. (1995). Interpretive case-studies in IS research – nature and method. *European Journal of Information Systems*, 4(2), 74–81.
Yeow, J., Edler, J. (2012). Innovation procurement as projects. *Journal of Public Procurement*, 12(4), 488–520.
Yin, R. (1994). *Case Study Research: Design and Methods*, 2nd edn. Thousand Oaks, CA: Sage.
Yin, R.K. (2003). *Applications of Case Study Research*, 2nd edn. Thousand Oaks, CA, USA and London, UK: Sage Publications.
Zheng, J., Knight, L., Harland, C., Humby, S., James, K. (2007). An

analysis of research into the future of purchasing and supply management. *Journal of Purchasing and Supply Management*, 13(1), 69–83.
Zheng, J., Roehrich, J.K., Lewis, M.A. (2008). The dynamics of contractual and relational governance: evidence from long-term public–private procurement arrangements. *Journal of Purchasing and Supply Management*, 14(1), 43–54.

10. Public procurement for innovation in developing countries: the case of Petrobras

Cássio Garcia Ribeiro and André Tosi Furtado

INTRODUCTION

The public sector is a major user of many goods and services that support the functioning of the machinery of state and help meet the basic needs of the population. Government procurement plays a leading role in the promotion of domestic industry, especially given the magnitude of this market. Some research shows that government procurement accounts for 10–16 per cent of gross domestic product (GDP) in developed countries (Hoekman and Mavroidis, 1995; Audet, 2002; Georghiou et al., 2003; Weiss and Thurbon, 2006). Hence, in light of the scope and scale of the government procurement market, national governments use their purchasing power to promote the development of domestic productive sectors.

Government procurement policy has been at the centre of recent debate on policies to support innovation (Aschoff and Sofka, 2009). However, there is a lack of research on the adoption of government procurement policy as an instrument to stimulate innovation in developing countries (Ribeiro, 2009). This chapter sets out to start filling this gap by attempting to elucidate the role that can be played by government procurement policy in fostering innovation by firms located in developing countries.

The chapter focuses on a specific policy instrument, which is public procurement for innovation (PPI) (Edquist and Zabala-Iturriagagoitia, 2012).[1] In contrast to off-the-shelf purchasing of standardized goods and services, PPI is characterized by the acquisition of goods and services that constitute innovations, but did not exist at the time the procurement call was launched.

PPI policies have been decisive for the technological capability-building of equipment and service suppliers in several developed countries (Gregersen, 1992; Fridlund, 1993; Edquist and Hommen, 1998; Llerena et al., 2000). In the oil industry there are important cases of successful procurement policies in which state companies are used to promote their own oil supplies industry, such as in France or in Norway (Cook, 1985; Furtado, 1994).

It is important to note that the concept of innovation used here encompasses a range of activities from copying, imitation, enhancement and experimentation to more value-added activities such as the development of new products and processes. This extended view of innovation, which transcends the traditional concept, is particularly important for an assessment of technical change in developing countries (Dosi, 1988; Lall, 1992; Bell and Pavitt, 1993). In developing countries, firms are recipients rather than developers of technology, and the learning process is the main way in which they build up technological capabilities (Lall, 1980, 1982, 1992; Bell, 1984). Taking into account these specificities, this chapter aims to analyse whether PPI in developing countries can promote innovation in firms, and also to understand the practical limits of such policy. To attain this goal, the chapter provides a conceptual framework to classify and evaluate the innovative impacts of PPI in developing countries.

The case of Petrobras in Brazil is examined here. In particular, the chapter addresses the procurement policy for the offshore projects of this state-owned enterprise.[2] Despite being from a developing country, Petrobras is recognized as a company at the technological cutting edge, with respect to the exploration and production of oil and gas (O&G) in deep waters (Furtado, 1999). Since its foundation in 1953, Petrobras has pursued a procurement policy that is very important for the constitution of the Brazilian oil supply industry[3] (BOSI). The creation of Petrobras is associated with Brazil's heavy industrialization.[4] One of the main challenges of the country's development was related to the necessity to attain self-sufficiency in oil, because its imports were becoming an increasing burden for the national trade balance. The efforts to build the Brazilian oil refineries and to expand national oil production required heavy investments in equipment that was not produced locally at the time. This is why Petrobras has engaged in the promotion of the BOSI since the middle of the last century (Macedo e Silva, 1985). The Petrobras procurement policy became an example for other Brazilian state companies during the 1970s, such as Eletrobras, Siderbras and Telebras, which intended to adopt similar local content policies (Villela, 1984). The Petrobras local procurement policy was also improved in the second half of the 1970s and in the 1980s, the purpose being to internalize the offshore production systems (Surrey, 1987).

In the Brazilian case there is no clear frontier between federal government and Petrobras. The state company itself has taken on the task of attaining oil self-sufficiency and promoting the capital goods and engineering sectors (Villela, 1984; Alveal Contreras, 1994; Freitas, 1999). Thus, it is an interesting case for an investigation of whether public procurement policy in developing countries is used as a stimulus to innovation. To elucidate the impact of Petrobras's procurement policy, this chapter proposes a case study approach. As argued in this book, to assess whether the procurement policy of a government entity may or may not be classified as innovative it is necessary to select an important case and investigate it thoroughly. Only through a systematic and micro-level approach is it possible to capture how the procurement policy contributes to transforming the innovative capabilities of local firms.

This chapter focuses on large platforms and complex technological projects, and more specifically on the case of P-51 construction. This project is not only significant in its economic size (around US$1 billion) but also in its technological challenges. P-51 is one of the largest semi-submersible platforms[5] ever built in the world. As for the technological challenges, it is important to understand that the construction of a platform such as P-51 involves a large number of engineering-intensive activities with custom-built items and complex interfaces. Moreover, this stationary production unit (SPU) is the first platform of its type built entirely in Brazil. In sum, the project is emblematic in several ways.

Because of its importance, the P-51 platform can be considered as a case of Petrobras's PPI for the BOSI. Besides, the acquisitions made by this company represent an important part of the government procurement market in Brazil. Last but not least, we consider that this analysis contributes to the understanding of some important features of PPI in developing countries.

The chapter presents the findings of a field survey covering all the key players involved in the P-51 project: Petrobras, its engineering, procurement and construction (EPC) contractors,[6] and a sample of subcontracted suppliers. The results of the field survey suggest that participation in the design of P-51 led to a small amount of innovation learning experience for the main local firms involved. According to information obtained from interviews, the main obstacles to Petrobras's PPI were that the main multinationals participating in the project did not conduct high-tech activities in Brazil, and that Brazilian firms had a low level of technological capability.

The chapter has six sections including this introduction. The next section analyses government procurement policy, emphasizing PPI. The focus of the third section is the literature on technological learning and the typology created to evaluate PPI in developing countries. Next, Petrobras's procurement policy is discussed. The key features and procurement of P-51 are outlined in the fifth section, which goes on to describe the model proposed in this chapter for evaluating the learning experiences made possible by participation in oil production projects, before presenting the empirical findings of the study conducted to analyse P-51 procurement. The last section contains the conclusions.

PUBLIC PROCUREMENT FOR INNOVATION

Goods and services are purchased by the public sector to supply the inputs required for the fulfillment of its functions, such as the provision of public goods for use by society. In addition, government procurement policy is one of the few industrial policy instruments still in use by national governments to promote the development of the domestic industry, given the existence of protectionist safeguards in the multilateral rules that discipline this field (Audet, 2002).

Although the supranational rules applicable to this field are designed to liberalize the market, it can be argued that not all markets will be opened up to international competition, since areas such as defence spending and, to some extent, investment in research tend to be reserved for domestic suppliers. Moreover, while purchasing decisions by private players are guided by strictly market-based criteria such as price, lead time and quality, public sector agencies can use other criteria, depending on the industrial policy and science, technology and innovation (ST&I) agenda adopted by the national government (Ribeiro, 2009).

Edquist and Hommen (1998, 2000) consider the idea that government procurement policy may be not just a modality of industrial policy but also an instrument for stimulating innovation. They emphasize the linkages between government procurement policy and innovation theory, highlighting the public sector purchases that contribute to the development of user–supplier interaction. For Edquist and Hommen (1998), two types of government procurement can be identified: regular government procurement, and public procurement for innovation (PPI). Regular government procurement occurs when a government agency buys standardized goods, such as pens and paper. PPI occurs 'when a government agency places an order for a product or system which does not exist at the time, but which could (probably) be developed within a reasonable period' (Edquist and Hommen, 1998, p. 4).

Geroski (1990) propounds the view that PPI is a more efficient instrument for fostering innovation than a wide array of R&D incentives, the tool used most frequently by national governments to promote domestic business innovation. When the end-user of the procured goods is the public sector itself,[7] the user–supplier relationship is of a special type, insofar as the final customer has substantial purchasing power and in some cases cutting-edge technological capabilities (Ribeiro, 2005). It is worth noting that innovative projects often fail to materialize owing to the risks and large amount of capital they usually involve. Thus the goods, processes, services and systems generated by government orders might not be developed at all if that order did not exist.

Despite the importance of PPI to the technological development of a country's industrial fabric, it is possible to identify cases where the end result of such a policy has been failure, in technological and/or commercial terms. Examples include the US nuclear power industry (Campbell, 1988; Edquist and Hommen, 1998), and the Anglo-French Concorde programme (Edquist and Hommen, 1998). Notwithstanding such failures, there are also countless success stories.[8] Moreover, the risk of failure is a necessary element of any innovation project and it therefore cannot be said to invalidate the deployment of an innovative public procurement policy.

It should be stressed that many innovations associated with strong social needs have low internal return rates. In such cases, PPI also serves as an important driver of innovation development. Thus, considering that the market does not always offer effective mechanisms for the satisfaction of society's needs, PPI can help fill the gaps created by market failure, by fostering, for example, the production of innovative public goods (Dalpé, 1994).

The idea that government procurement policy can act as an important stimulus to innovation is grounded in arguments generally used to support public sector intervention in a country's technological development. Such arguments can be summarized as follows:

- Underinvestment in high-risk or long-term research and development (R&D), involving radical innovation.
- Underinvestment in socially desirable technologies, or overinvestment in socially undesirable technologies.
- Technological backwardness in some industries.
- Large-scale projects.
- Military requirements or needs in terms of economic security for domestic capacity-building in strategic technologies (Gregersen, 1992; Rolfstam, 2005).

Despite the importance of PPI as a government policy instrument, this means of satisfying demand has been overlooked or underestimated for many years (Edler and Georghiou, 2007). The literature on this subject is relatively recent and mostly restricted to the experience of developed countries. The main studies aim to present the achievements of PPI in the US (see, for example, Cohen and Noll, 1991) and Europe (Edquist et al., 2000; Edler et al., 2005). Thus, as stressed in the introduction, one of the main aims of this chapter is the presentation of elements that contribute to an understanding of innovative public procurement in the developing countries.

TECHNOLOGICAL LEARNING AND PPI IN DEVELOPING COUNTRIES

For a long time the economic literature has neglected questions relating to the technical progress achieved in developing countries. The basic neoclassical assumption is that technology is freely available to all countries and to all firms within each country. In this context, developing countries are assumed to prefer the technologies created by developed countries, selecting the most appropriate according to the factor–price ratio, in view of their physical, capital and labour endowments (Lall, 1994).

In turn, non-conventional theory has also neglected for a long time the role played by technological efforts in developing countries. Lall (1982) argues that technical change in developing-country industries differs from the Schumpeterian concept of radical innovation. The fact that technological activities by firms in these countries do not fit the Schumpeterian concept has delayed an examination of the changes that occurred during the industrialization process, thus concealing the richness and diversity of these experiences.

The 1970s is considered a watershed in the debate on developing countries' technological efforts. Here it is worth noting the importance of the groundbreaking ideas proposed by evolutionary theorists such as Nelson and Winter (1977). In their seminal work, any conscious technical change made by a firm to enable the introduction of new products and processes can be considered as technical progress, even when other firms know about the resulting technology. According to this vision, any innovation requires some kind of effort by the firm in question, regardless of whether it is added to the global stock of existing technologies, reproduces technology already developed by another firm, or even simplifies or renovates existing technology.

Inspired by this contribution from Nelson and Winter (1977), studies were produced on the experience of late-industrializing countries, showing that certain characteristics of externally acquired technology and of the markets in which technology is bought and sold, such as non-replicability and imperfect information, lead firms in these countries to undergo a process of technological learning with a strongly idiosyncratic and adaptive bias (Katz, 1974, 1981, 1987; Lall, 1978; Fransman, 1984). In this context, technological learning enables developing-country firms to acquire the necessary competencies to select, assimilate, adapt and improve imported technology.

According to Lall (1980), developing-country firms learn about foreign technology in three stages: elementary – operational learning and minor adaptations; intermediate – copying and enhancement of processes and products; and advanced – understanding integrated projects and innovation (Table 10.1). The following forms of learning are part of these stages.

Table 10.1 Learning modes by developing-country firms

Stage	Learning mode
Elementary stage	Learning by doing: involves the development of manufacturing skills; materialized in manufacturing skills, promoting the development of increasing skill in production so that labour costs per unit of output and the incidence of quality issues can be reduced. Learning by adapting: involves initiatives designed to obtain the mastery of know-how that enables the introduction of small adaptations to industrial plants or products.
Intermediate stage	Learning by design: involves the development of skills and knowledge relating to basic design; replication of external product or process technology by the firm; and hence the possibility of opening the black box of a specific technology to capture the knowledge essential to the redesigning of complex equipment. Learning by improved design: involves modification and/or improvement of the basic design; this is the second step in engineering projects, whereby changes are made to products, processes or industrial plants to adapt them to local specificities in terms of raw materials, conditions and skills.
Advanced stage	Learning by setting up complete production systems: involves the development of the technical expertise required to produce equipment, as well as engineering skills and adaptation of entire factories according to specific needs. Learning by innovation: involves the creation of new products and/or processes by late-industrializing countries with characteristics that differ from those found in the developed countries.

Source: Based on Lall (1980).

Notwithstanding the descriptive value and topicality of the classification proposed by Lall, the same classification is put forward here, in line with the broader view of innovation adopted in this chapter and with the technological profile of the firms that operate in developing countries. Thus we separate procurement learning into three groups of learning: low-, intermediate- and advanced-innovative learning.

Low-innovative learning promotes limited changes of the technologies with which the firm operates and manufactures, involving a combination of factors such as skills, equipment, product and production specifications, organizational systems and methods, all of which are usual for the firm. This kind of learning develops skills in the manufacturing process, enabling the incidence of quality issues to be reduced, for example, or minor adaptations to be made to an industrial plant or to products. Based on this description and the concept of innovation adopted in this chapter, it can be inferred that learning of this kind promotes limited innovativeness in firms. Lall (1980) classifies it as elementary learning, including as part of this category learning by doing and learning by adapting.

Intermediate-innovative learning enables the firm to develop the technological capabilities required to copy products and processes, make incremental improvements and develop new ones by incorporating the additional resources needed to generate and manage innovative technological activities. The intermediate kind of learning is related to adaptative PPI, in which the new product or system is incremental and new only to the country (Edquist and Zabala-Iturriagagoitia, 2012). Thus intermediate-innovative learning includes learning by design, learning by improved design, learning by setting up a complete production system, and learning by innovation.

Advanced-innovative learning is associated with the generation and commercialization of new technologies. This kind of learning involves ordering of items that require the performance of risky and highly complex activities by the suppliers and often involve large investments in R&D. This kind of public acquisition can promote the development of technical skills required to produce items of equipment involving engineering skills and adaptation of entire factories or plants according to specific needs (learning by setting up a complete production system), as well as the emergence of new products and processes (learning by innovation). This kind of learning is associated with developmental PPI that implies new-to-the-world products or systems (Edquist and Zabala-Iturriagagoitia, 2012).

This classification is used to propose a typology for evaluating the impact of government procurement policy on local suppliers, focusing on the case of developing countries. The main modification is that we propose an additional level of public procurement related to elementary and routine learning that is

Table 10.2 Typology of government procurement for developing countries

Characterization of activities	Low-innovative	Intermediate-innovative	Advanced-innovative
	Elementary; routine; experience-based	Intermediate; replicable; search-based	Advanced; inventive; R&D-based
Types of learning	Learning by doing; learning by adapting	Learning by design; learning by improved design	Learning by setting up a complete production system; learning by innovation
Procurement policy	Low innovative PPI	Intermediate innovative PPI (adaptative PPI)	Advanced innovative PPI (developmental PPI)

named low-innovative PPI. As can be seen in Table 10.2, government procurement is classified in accordance with the profile of the activities performed by firms engaged as contractors to the public sector and the associated learning mechanisms.

The typology presented in Table 10.2 shows the different ways of technological learning that can be triggered from a government order. With this approach, we seek to escape the non-innovative/innovative dichotomy (which could lead to misinterpretation of what was framed as 'innovative', considering only advanced activities, based on R&D), taking into account the diverse learning opportunities engendered by participation in a government order.

This typology emphasizes the importance of different learning processes in firms located in developing countries, and addresses the multiplicity of activities from which technical change is achieved; it is, therefore, more adherent to developing countries' reality. It is evident that the adoption of advanced PPI appears more promising in terms of fostering suppliers' technological capabilities. However, one should not disregard the importance, for the technological reality of developing countries, of public acquisitions that promote low-innovative and intermediate-innovative learning. Intermediate-innovative learning is important for technical change that takes place in these countries, because it involves more deliberate effort (than in the case of low-innovative learning) by the public sector's suppliers, combined with activities such as copying and/or improvement of licensed technologies or improvements in specifications given; stimulating, therefore, the capability of these firms in terms of basic engineering.

THE EVOLUTION OF PETROBRAS'S PROCUREMENT POLICY

Petrobras is a state-owned enterprise classified under Brazilian law as a 'mixed-capital' corporation because a large proportion of its capital is held by private investors, but the majority of its voting share is under the control of the federal government of Brazil.[9] This company can be considered the first Brazilian state-owned enterprise to adopt a procurement policy in which the domestic supply is favoured. Little more than a decade after its foundation, Petrobras unveiled the Gabriel Passos and Alberto Pasqualini refineries with 80 per cent local content. However, during the import substitution period its procurement policy failed in some respects, such as not giving enough importance to the promotion of the BOSI's innovative technological learning (Ribeiro, 2005).

In the second half of the 1990s, Brazilian policy-makers removed government procurement from their priorities and instead advocated a new type of state intervention in economic affairs. Belief in the 'invisible hand' of the market and the negative effects of the 'hypertrophy' of the state led to deep-seated changes in public sector involvement in the economy, especially privatization, deregulation, trade liberalization and the adoption of orthodox macroeconomic policies to stabilize the currency (Ribeiro, 2005).

As part of this process, Petrobras lost its monopoly on oil extraction, refining, transport and distribution. The new political, economic and regulatory framework gave more autonomy to the state company. This increasing autonomy, associated with the liberal policy orientations and with the structural changes that were happening in industry at the world level, induced Petrobras to reduce local procurement, which had adverse effects on the BOSI. The procurement policy of Petrobras was associated with a high level of engagement of this company in the design and engineering of major investments projects. However, in line with global trends in the oil and gas industry, Petrobras began outsourcing many of the functions it had previously undertaken in these projects, preferring to engage turnkey contractors. Foreign firms were better equipped to act as EPC contractors in these projects, and local procurement declined further because the foreign firms were not interested in building ties with the BOSI (Furtado et al., 2003).

Since the beginning of the 2000s, however, it has been possible to perceive a redirecting of Petrobras's procurement policy back toward the domestic market, as evidenced by the adoption of local content requirements in competitive bidding procedures for offshore platform projects. The case of P-51 is emblematic of this turning point of the Petrobras procurement policy. The local content required for the topside was 60 per cent, while for the generation and compression modules it was 75 per cent, excluding critical equipment such as turbines and compressors (Furtado et al., 2003). The hull, which at the beginning[10] was excluded from the local content requirements, was afterwards included in the local content requirements, and entirely built in Brazil. The case of P-51 is emblematic because it is the first semi-submersible platform assembled entirely in Brazil (Ribeiro, 2009).

Having indicated the significance of P-51 in the context of Petrobras's new procurement policy, we proceed in the next section to presenting the findings of a case study focusing on P-51. This policy however was much more oriented towards local content and job creation than to improving local technological capabilities. Our main research question lies in identifying how this kind of public procurement can also promote innovation. Semi-structured interviews with managers of Petrobras, EPC contractors and a sample of subcontractors for P-51 were used as the main data collection tool for the field survey. Interviews took place in 2007 and 2009. The main aim of the field survey was to find out whether participation in the P-51 project was seen by EPC contractors and subcontractors as a good opportunity for technological learning in Brazil.

THE P-51 PLATFORM

P-51 is a semi-submersible platform that started to operate in January 2009. It is located in Module 2 of the Marlim Sul field in the Campos Basin offshore Rio de Janeiro state.[11] Constructed in modular fashion, P-51 comprises a hull and topside, with the latter containing all facilities and modules for utilities, processing, power, compression and accommodation. The entire platform weighs more than 40 000 tonnes, with a height of over 75 metres and a perimeter of 470 metres. It generates 100 Megawatts of electricity and can operate in water depths of up to 1800 metres. It is designed to function 24 hours a day for 25 years. Nominal throughput is up to 180 000 barrels per day, besides compression of 6 million cubic metres of gas per day. Its accommodation module can house a crew of 200 (Reis, 2004).

The leadership of the P-51 project development process was UN-Rio (Petrobras's unit of exploration and production, E&P). In the capacity of 'end customer', UN-Rio elaborated the basis of the P-51, designing a descriptive document with some basic features and specifications of the platform. This document was sent to the Leopoldo Américo Miguez de Mello Research and Development Center (CENPES) which, based on this initial input, was in charge of developing a basic design for the P-51 platform. The basic design developed by CENPES encompassed small line drawings, approximate dimensions, some details of the process plant and simulations on the basis of oil, aiming primarily to set out the size of equipment and thus help in defining the size of the unit.

The basic design subsidized activities performed by contractors for the front end engineering design (FEED) execution. FEED is an intermediate stage between basic and detail engineering, comprising the production of drawings, specification of steel weight, and design of mission-critical equipment such as compressors and generator groups. Petrobras formerly used the basic design in the bidding process for hiring their platforms, which did not contain the list of materials, weight and detailing about the equipment used. FEED represents an effort to fill these gaps, without reaching the level of detail of the executive project, but allowing the preparation of a list of materials and equipment. With FEED at hand and assisting greatly in the commercial evaluation of tenders, Petrobras has greater confidence with respect to the budgeting of the project. FEED also allows the market to make proposals that are neither too low nor absurdly high, since the lists of documents generated in this activity were used in the bidding process of P-51. The quotations made by the companies that participated in the bidding are based on such lists, as are proposals made by them.

Upon the conclusion of P-51's basic design, Petrobras placed a bid for FEED, in which event the winning companies were Aker from Norway, and UTC from Brazil. The Norwegian company was contracted to perform FEED for P-51's hull, while the Brazilian company was hired to carry out the topside FEED. CENPES participated in the realization of these two projects. Some activities, such as specific calculations, were transferred to CENPES.

The pre-bidding of P-51 followed the sequence described above. Once this first step was completed, a bidding process for the effective construction of this oil platform was opened. The platform construction involved three major contracts: a contract for the construction and delivery of natural gas compression modules; a contract for the construction and delivery of power generation modules; and a contract for hull construction and integration, topside construction, process plant construction, and reception and integration of all modules.

According to information obtained in the interviews, Petrobras could hire only one company for the execution of all such contracts. However, dividing the project into modules and purchasing them from different companies allowed the manufacture of large machines (like the turbogenerators) in a separate manner. By modularizing the project, all parts of the platform have the same starting point and are manufactured simultaneously, so as to allow time for supplying the modules, and reducing the risk of delivery delay. To avoid delays in the execution of the work and any other problems, Petrobras allowed the formation of consortia.

Eleven companies were invited to participate in the bidding process of P-51 (by letter of invitation). After the result of the bidding, Petrobras invalidated the process because of the high prices of the proposals, and went through a specific negotiation with pre-qualified companies. At the same time as this negotiation, three contracts were signed in 2004 with three different EPC contractors:[12] a contract for the construction and delivery of natural gas compression modules (Rolls-Royce[13]); a contract for the construction and delivery of power generation modules (Nuovo Pignone[14]); and a contract for hull construction and integration, topside construction, process plant construction, and receipt and integration of all modules (Fels Setal Technip

Consortium – FSTP[15]). Construction of P-51 took place from 2005 to 2008 (Ribeiro, 2009).

The contracts of the P-51 were executed in turnkey lump sum modality (global enterprise with fixed price). In this modality the contractors were responsible for performing the engineering, procurement and construction (turnkey) following the design specifications determined by Petrobras–CENPES, and assumed responsibility for variations in project costs (lump sum). They also assumed the risks associated with project execution costs after the FEED endorsement,[16] which basically means they bore the cost of any budgeting and planning errors once they had technically endorsed the design.

P-51 is a complex production system with a series of custom-built items (Barlow, 2000). It is arguable that participation in the project could lead to a significant amount of innovative learning, especially for EPC contractors and suppliers of mission-critical items. Besides, this is an emblematic case about the recent change in Petrobras's procurement policy, characterized by local content requirements. These facts evidence the importance of a study that establishes the linkages between the procurement policy adopted by Petrobras for the P-51 order and the technological learning opportunities provided by participation in the project for EPC contractors and local subcontractors.

Technological Learning in Platform Projects

The various types of players involved in any oil platform project perform functions and activities with different levels of technological complexity, which can lead to low-innovative, intermediate-innovative or advanced-innovative learning. Table 10.3 describes the range of possibilities for technological learning that are derived from a firm's participation in an oil platform project, and helps to interpret the effects of the procurement policy adopted by Petrobras.

The matrix of technological learning in platform projects presented here is an adaptation of the technological capabilities matrix developed by Lall (1992), later adapted by Figueiredo for empirical use in studies of steel companies (Figueiredo, 2002, 2004) and organizations that operate on the basis of information and communications technology (Figueiredo, 2006). But, unlike Lall's matrix and the adaptation proposed by Figueiredo, our approach evaluates technological learning arising from

Table 10.3 Technological learning in platform projects

Learning →	Level of learning		
	Low-innovative	Intermediate-innovative	Advanced-innovative
Functions ↓	Routine; experience-based	Replicable; search-based	Inventive; R&D-based
Platform engineering design	Detailed and routine design (learning by doing)	FEED (learning by improved design); FEED endorsement (learning by improved design)	In-house production of basic design (learning by innovation)
Fabrication of equipment and supply of services	Replication of given specifications (learning by doing); routine quality control to assure conformity with customer's specifications and standards (learning by doing); minor adaptations to given specifications (learning by adapting)	Improvements to given specifications (learning by improved design); reverse engineering (learning by design)	Development of entirely new equipment or services (learning by innovation)
Construction and assembly	Fabrication of hull or topside modules in conformity with given specifications (learning by doing); minor adaptations to hull or topside fabrication (learning by adapting)	Modifications to hull or topside fabrication (learning by improved design); modifications to module integration or deck mating (learning by improved design)	Development of new platform construction and assembly methods/solutions via engineering and R&D (learning by innovation); EPC management of platform projects (learning by setting up complete production system)

Source: Adapted from Lall (1992) and Figueiredo (2002).

participation in a project and not the technological capacity that the company already has.

The matrix covers the following functions (rows): platform engineering design, fabrication of equipment and supply of services, and construction and assembly. To each of these functions correspond activities that may involve low-innovative, intermediate-innovative or advanced-innovative learning (columns). To create this matrix, the selection of key technological functions in platform design and construction was necessary, as was the identification of specific activities that offer opportunities for learning in the various forms outlined in the section on 'Public procurement for innovation'.

Low innovative learning promotes knowledge about the technologies operated by the company supplying the public sector, and involves a combination of factors such as skills, equipment, product specifications and production, systems and organizational methods with which it is associated. Intermediate innovative learning is linked to activities that promote technological capability in the suppliers to copy and promote improvements or adaptations in specific technologies. Finally, advanced innovative learning involves ordering of items that require the performance of risky and highly complex activities by the suppliers, as well as large investments in R&D.

Based on these definitions and on the list of activities related to the three functions performed by the suppliers of an oil platform (engineering design, fabrication of equipment and supply of services, and construction and assembly), Lall's modalities of learning presented in the section on 'Technological learning and PPI in developing countries' were adapted to construct the matrix of technological learning in oil platform projects. This model assumes that the study of the technological learning stage (basic, intermediate or advanced) brings important contributions to the analysis of the profile of procurement policy, particularly in the case of developing countries, and, more than that, reveals the level of technological capability of the respective companies.

As pointed out in the introduction, the procurement of a platform such as P-51 involves a large number of engineering-intensive activities with custom-built items and complex interfaces. Moreover, this is the first platform of its type built entirely in Brazil. As a result, we can say that P-51, in theory, represented a good opportunity for the BOSI's companies not just in terms of low-innovative, but also mainly intermediate-innovative and advanced-innovative modes of learning.

It should be noted that this model was validated during the field survey with different professionals from the Brazilian oil industry. Furthermore, it is noteworthy that, although this model has been applied to analyse an oil platform project, it can be adapted for studies related to other projects and industries.

Technological Learning in Brazil by EPC Contractors Working on P-51

The EPC contractors who worked on P-51 are analysed in this section in light of the matrix presented above. As noted, the analysis of participation by these companies in the platform project is based on information collected from interviews with their managers. Table 10.4 shows the functions and activities performed in Brazil by EPC contractors on P-51 and the respective forms of technological learning involved.

The point to which attention should be paid is the fact that only one of the EPC contractors directly performed the functions of platform engineering design and construction and assembly in Brazil. Rolls-Royce and Nuovo Pignone outsourced these functions for power generation and gas compression modules to firms and shipyards located in Brazil, which had the mission-critical components of their modules manufactured outside Brazil.

Rolls-Royce has a sales office and turbine maintenance and repair facility in Brazil. Nuovo Pignone only has a sales office there. Rolls-Royce produced the turbogenerators, the main components of the power generation modules for P-51, at its plant in Liverpool, UK. Likewise, Nuovo Pignone produced the main components of the compression modules (reciprocating compressors) abroad, more precisely in Florence, Italy. In sum, these firms did not directly perform any of the functions or activities for P-51 in Brazil, and hence are not included in Table 10.4.

The FSTP Consortium, the only EPC contractor performing some technological activities in Brazil, included the firms Keppel Fels and Technip. With regard to platform engineering design, as shown in Table 10.4, FSTP partially produced detail engineering for its work on P-51 in Brazil. 'Partially', because KeppelFELS produced FEED and detail engineering for the hull in Singapore, while Technip produced FEED and detail engineering for the

Table 10.4 EPC contractors: learning experience in P-51 project

Learning →	Level of learning		
	Low-innovative	Intermediate-innovative	Advanced-inventive
Functions ↓	Elementary; routine; experience-based	Intermediate; replicable; search-based	Advanced; inventive; R&D-based
Platform engineering design		Detail engineering of topside partially produced in Brazil (FSTP)	
Fabrication of equipment and supply of services			
Construction and assembly	Fabrication of process modules according to given specifications (FSTP); minor adaptations to fabrication of process modules (FSTP)	Modifications to hull fabrication (FSTP); modifications to topside construction (FSTP); modifications to deck mating (FSTP)	EPC management of platform projects (FSTP)

topside in France and Brazil. Detail engineering is based on FEED specifications and uses platform information of various kinds. The aim of this activity is to produce a set of documents with detail designs, drawings, bills of materials, specifications for facilities and materials, architectural designs and calculations, among other items. These documents then serve as key inputs for the procurement of equipment, services and materials, and for platform construction.

With regard to the production of equipment and supply of services, FSTP farmed this function out to local and foreign firms and therefore did not engage in any learning experiences associated with it. The activities performed by FSTP in Brazil were confined to hull and process module construction and assembly, as well as integration of the modules and deck mating. Hull fabrication was a new activity for the consortium in Brazil. They were received by the BrasFELS shipyard and welded together to form two C-shaped structures. The two C-shaped structures were joined at sea in an operation without precedent in Brazilian naval engineering, giving rise to a floating ring on which four columns were anchored to serve as the foundations for the topside.

Another point stressed in the interviews with regard to the construction activities performed by the consortium was the need to use special materials to build the process plant, given the weight, viscosity and corrosiveness of the oil processed by P-51. These characteristics of the oil extracted from Module 2 by Marlim Sul required the use of special stainless steel called super duplex. Welders at the KeppelFELS shipyard were obliged to adapt the procedure used to weld regular steel for oil rigs in order to guarantee the quality of super duplex steel welds.

Also, with regard to the activities performed by FSTP on P-51, fabricating the topside outside the barge used for deck mating was a significant challenge. The FS-1 barge, built by the BrasFELS shipyard for work on P-52 as well as P-51, was being used for the construction of P-52 when work began on P-51.[17] The solution was to fabricate the topside deckbox on one of the slipways located just behind the shipyard's dry dock[18] and then skid the finished deckbox down the slipway to the hull. A 112 metre slipway was built to slide the deckbox down on to the barge. First the dry dock was opened up to receive the FS-1; then, once the dry dock had been drained, the deckbox was skidded to the barge using hydraulic jacks.

Finally, with regard to the deck mating operation, although FSTP had already executed a similar operation for P-52 in Brazil, it can be said that a modification was introduced for the P-51 deck mating operation: installation of the flare boom[19] on the topside before this structure was mated with the lower hull. This adaptation to the platform assembly process resulted in valuable cost and time savings by dispensing with the use of barges for hoisting and installation of the flare boom after deck mating. Among the hazards avoided by this solution was the risk of topside listing or tilting, since most of the flare boom projects were outside the structure. According to information collected in the field survey, the operation was a success.

Cross-referencing the information collected from FSTP with the learning matrix, presented in the previous section (see Table 10.1), shows that the activities performed by the consortium for the P-51 project led to both innovative and non-innovative learning. Learning by doing took place in the fabrication of the process modules in accordance with the FEED specifications. Learning by adapting consisted of the introduction of a minor adaptation to fabrication of the process module by using special steel that is difficult to weld. Learning by improved design derived from topside detail engineering and the modifications adopted for construction of the hull and topside and for the deck mating operation. Lastly, based on the information obtained in the field survey it can be said that FSTP, as the lead EPC contractor for P-51, benefited from learning by setting up a complete production system, as it managed both the P-51 and P-52 projects under the EPC regime.

Seven occurrences of learning in Brazil by FSTP can therefore be identified. In this sense, it can be said that the balance of innovation for this EPC contractor was positive in the case of P-51. However, considering that the other two EPC contractors did not experience any kind of learning in Brazil in connection with the platform project analysed in this chapter, it can be concluded that the total innovation impact of this Petrobras order on EPC contractors was small.

Technological Learning in Brazil by Subcontractors Working on P-51

The activities of local firms subcontracted by the EPC contractors to supply equipment and services for the P-51 project were also investigated in the field survey, which included a sample of 12 such firms. The subcontractors that are presented in Table 10.5 belong to the BOSI.

The companies were selected for sampling purposes according to the following criteria: (1) located in Brazil (regardless of whether the parent company was Brazilian); and (2) supplier of an important item for P-51 in the eyes of EPC contractors and Petrobras. The sample encompasses the main firms involved in hull block construction and in construction and assembly of metal structures for the power generation and compression modules, as well as some of the local firms subcontracted to supply mission-critical equipment and services for the platform (such as pressure vessels, electricity generators and the electrical system), and the firms responsible for detail engineering of four out of the platform's eight modules.

The firms were divided into three groups according to the activities performed and the items supplied, namely: construction and assembly (firms 1 and 2); equipment suppliers (3, 4, 5, 6, 7, 8, 9 and 10); and engineering firms (11 and 12). These firms were mostly located in São Paulo state and Rio de Janeiro state, the wealthiest state in Brazil and the state leader in oil and gas production.

Eight of the 12 firms (about 65 per cent) can be classed as large according to the criteria used by Instituto Brasileiro de Geografia e Estatística (IBGE), the national statistics bureau.[20] Thus for most of the firms size was not a factor that might have restricted their technological dynamism. Moreover, 75 per cent are Brazilian-owned, so that origin of capital cannot in principle be used as an argument for failure to perform technological activities in Brazil.

It is worth noting that Petrobras considers the equipment and services provided by firms that are included in its Corporate Register as essential for the full operation of its plants. Therefore, we can say that all companies in the sample, to a greater or lesser extent, provide critical items for Petrobras's projects. The research revealed that the requirements of the Brazilian state-owned enterprise force BOSI's companies to improve their standards of quality, security, and health, safety and environment (HSE) and financial management, and so on.

Based on information collected in interviews, it was found that about 85 per cent (20 out of 23) of all occurrences of learning identified between subcontractors as a result of their

Table 10.5 Sample of P-51 subcontractors covered by field survey

Group	Firm	Location	No. of employees (2008)	Origin of Capital	Supplied to P-51
Construction and assembly	1	Itaguaí – RJ	1 200	Domestic	Hull block construction and assembly
	2	Ipatinga – MG	8 587	Domestic	Construction and assembly of metal structures for electricity generation and gas compression modules
Equipment	3	Araraquara – SP	2 814	Domestic	Pressure vessels
	4	Rio de Janeiro – RJ	637	Domestic	Pressure vessels, tanks, crane pedestal, lattice tubular structure for flare boom
	5	Jaraguá do Sul – SC	21 846	Domestic	Electricity generators
	6	São Gonçalo – RJ	60	Domestic	Air filter for gas turbine
	7	Jundiaí – SP	857	Foreign	Pumps and pumping systems in general
	8	Jundiaí – SP	9 030	Foreign	Electric system
	9	São Paulo – SP	250	Domestic	Control centre system, fire/gas detection panels
	10	Curitiba – PR	320	Domestic	Electromechanical assembly of power generation module
Engineering services	11	Rio de Janeiro – RJ	350	Domestic	Detail engineering for compression modules
	12	Macaé – RJ	880	Foreign	Detail engineering for power generation modules

participation in the design of the P-51 can be classified as basic-level (see Table 10.6).

Table 10.6 shows that low-innovative learning by doing was the most frequent mode of learning between the subcontracted suppliers in the sample, and was related to the following activities: replication of given specifications in equipment fabrication; routine quality control to assure conformity with Petrobras's specifications and standards; minor adaptations to given specifications; minor adaptations to hull block fabrication in accordance with specifications; and construction and assembly of metal structures for generation and compression modules according to given specifications.

The occurrences of intermediate learning accounted for about 13 per cent of total cases of learning observed among the sampled firms that participated as suppliers in P-51. They are related to the platform engineering design function (two), and the fabrication of equipment and supply of services function (one), in activities that enabled the firms involved to internalize learning by improved design. Table 10.6 clearly shows that none of the subcontractors experienced risky, R&D-based learning that could enable them to develop innovations at the international technology frontier.

Thus innovative activities were not the spotlight of the P-51 project. The almost total absence of technological efforts among the firms surveyed is one of the main factors contributing to the low technological dynamism of these firms. Only three of the sampled firms (25 per cent) reported the systematic activities and the existence of a formalized R&D department. Besides showing weak R&D investments, the information collected in the interviews suggests that these firms have negligible relationships with universities or research institutions.

The low level of technological effort in the BOSI and its weak involvement with other players in the national innovation system made the licensing of foreign technology inevitable. The study of the P-51 project revealed that most subcontractors relied heavily on technology licensing agreements or design development by the customer (Petrobras–CENPES) or by EPC contractors. The licensing strategy mitigated risk but prevented them from developing technological capabilities through more sophisticated forms of learning than learning by doing and learning by adapting.

Companies wishing to supply equipment and services to

Table 10.6 Subcontractors: learning experience in P-51 project

	Level of learning		
Learning →	Low-innovative	Intermediate-innovative	Advanced-innovative
Functions ↓	Elementary; routine; experience-based	Intermediate; replicable; search-based	Advanced; inventive; R&D-based
Platform engineering design		FEED/detail engineering for compression modules (firm 11) FEED/detail engineering for compression modules (firm 12)	
Fabrication of equipment and supply of services	Replication of given specifications in equipment fabrication (firms 3, 4, 5, 6, 7, 8, 9, 10); routine quality control to maintain existing specifications and standards (firms 3, 4, 5, 6, 7, 8, 9, 10); minor adaptations to specifications (firms 8, 9)	Improvements to specifications (firm 9)	
Construction and assembly	Minor adaptations to hull fabrication (firm 1); construction and assembly of metal structures for generation/compression modules according to given specifications (firm 2)		

Petrobras's projects need to be entered in the Corporate Register of Goods and Services Suppliers. This register is composed of qualified and enabled firms that meet the provisions of Decree 2.745/98, which governs the acquisitions made by Petrobras. The Brazilian state-owned enterprise evaluates companies seeking membership in the Corporate Register, based on some criteria, namely: technical, economic, legal, HSE, managerial and social responsibility. The aim of this rigorous selection is to ensure the quality of equipment and services used in its projects. Companies that seek registration by Petrobras are audited in the areas of engineering, procurement and manufacturing. In some cases, when applicable, they must develop prototypes and have the following certifications: ISO 9001, ISO 14000 and OHSAS 18000. The supplier that passes through Petrobras's sieve is incorporated into the Corporate Register and receives a Certificate of Registration of Registry Classification (CRCC), valid for one year.

In this regard, it should be noted that Petrobras has, besides the CRCC, a register called Brazilian Prime Vendor, composed of Brazilian companies that can provide equipment with third-party technology, since they own the domain of the manufacturing process. Firm 4, for example, by being a member of the Brazilian Prime Vendor register, acted this way in the projects of the P-55 and P-57. According to information obtained from firm 4, when local suppliers are contracted directly by Petrobras, the technology associated with the ordered item must be licensed by them, and this process is not accompanied by technology transfer. Thus, the technology is delivered by the company that holds it, to the domestic manufacturer, inside a black box.

Regarding the technological gap of BOSI, it is appropriate to present some information collected in the interview at firm 3. This firm provided three pressure vessels for process modules of the P-51, whose internal components were purchased outside of the country, more precisely from the Norwegian Aker Kvaerner. While firm 3 carried out activities related to industrial boilers, such as the manufacture of pressure vessels, Aker Kvaerner has focused on defining the internal components, critical to implementation of the physical and chemical processes within the platform. According to the interviewee from firm 3, this case serves to illustrate the technological dependence of the BOSI.

As stressed in the section on 'The evolution of Petrobras's procurement policy', one of the problems identified by the literature in relation to Petrobras's procurement policy during the import substitution industrialization period refers to the fact that this policy has not dealt with the BOSI's technological capacity to innovate. More recent studies indicate that this scenario has not changed lately. According to Dantas (2004), the lack of technological capability and the dependence on foreign technologies are problems that remain from the import substitution industrialization period. Oliveira and Rocha (2008) state that even companies with higher competitive capacity in this industry are dependent on external authorization and licensing of the technologies they use. Thus, these authors conclude that the BOSI presents a significant deficit of competitiveness linked to the fragility of national engineering and limited technological capacity to innovate. Ruas (2010, p. 41) states that despite the growth experienced by the BOSI in recent years, due to the increased volume of orders from Petrobras, 'some deficiencies persist, limiting the potential for advances even with the advent of reserves and production in pre-salt layer'.

The results of the research presented in this chapter converge with these studies, revealing that the low technological capacity of the BOSI is the main obstacle to its development. Therefore, the simple replication of externally developed technologies leads mainly to low innovative learning. Despite its importance, it is the first stage of learning in the classification proposed by Lall (1982). The technological development of an industry requires that firms are able to move to more advanced stages in terms of learning. For such a development to take place, a more active attitude is required by firms receiving technology coming from abroad (Ribeiro, 2009). However, only three companies in the sample (25 per cent) carried out R&D systematically and had a formalized department for this purpose. Therefore, it is not surprising that there was no occurrence of advanced learning between the subcontracted companies of the sample, since such learning requires a more deliberate effort on the part of companies.

According to what was found in the interviews, Petrobras's procurement policy could also help to change this picture. The technological capabilities accumulated by CENPES own the domain of much of the technology used in the design of a platform. This domain comprises the basic engineering, responsible for defining the size and model of the platform, equipment and materials used, the operating conditions of the processing fluids extracted from wells, and so on. However, the CENPES technological domain is not transferred to Brazilian suppliers. In accordance with the field research, what happens in relation to the basic engineering of platforms is symptomatic. For this activity, CENPES, according to some respondents, is the main competitor of engineering companies in the country. This competition prevents the learning of these companies in basic engineering of platforms, which is restricted to detailed engineering.

From this study it was also possible to verify that the links between EPC contractors and subcontractors are purely commercial. From the field research, we can say that the EPC's concern is to acquire equipment or services from a company with CRCC. An indication of the fragile relationship between EPC contractors and subcontractors is the absence of clauses in the contracts between the parties providing the transfer of technologies and conducting training.

Thus, taking the case of P-51 as an illustration of the procurement policy adopted by Petrobras for its offshore projects in recent years, it can be said that the national operator redirected its purchasing to the domestic market and thereby contributed significantly to a recovery by important segments of the BOSI, such as shipbuilding and naval engineering. Nevertheless, its procurement policy did not advance in the sense of reducing this industry's technological dependency, because it did not foster the technological learning most, which made it possible to perform an authentic process of catching up.

CONCLUSIONS

Procurement by government entities can play a leading role in the promotion of domestic industries, above all owing to the size of this market in the national economy. This chapter aims to contribute to the debate about procurement policy by focusing on its consequences in terms of technological learning and innovativeness in developing countries.

Developing-country firms are much more recipients than developers of technology, so that the learning process is the principal means through which they can build technological capacity. The chapter therefore presents a methodology for analysis of public procurement policy in developing countries, drawing from the literature on technological learning. The thesis presented here is that this type of approach is a useful tool for determining the extent to which public sector procurement stimulates innovation by firms located in developing countries.

The chapter uses this approach to analyse the case of Petrobras, chosen mainly because of its technological vigour in deepwater oil production. In addition, the use of procurement by Petrobras to reinvigorate domestic industry was put back on the Brazilian government's agenda at the beginning of the 2000s. The bidding and construction process for the P-51 platform is part of this context. These factors combined justify the choice of Petrobras and P-51 for application of the framework proposed in this chapter to evaluate the level of technological learning of the BOSI. To evaluate the impact of P-51 procurement on the BOSI, the chapter presents the findings of a field survey comprising interviews with the key players in the project, namely Petrobras, its EPC contractors and a sample of subcontracted suppliers.

For EPC contractors, the field survey showed that only one experienced some technological learning thanks to the P-51 project. The other contractors farmed out detail engineering, procurement of goods and services, and module construction to subcontractors. This shows a lack of interest of some multinationals in manufacturing and conducting R&D in Brazil, which has restrained technological learning by the BOSI.

Most of the critical, expensive and technology-intensive components were fabricated outside Brazil. While complying with the contractual local content requirement, it excluded some of the major high-value imported items. One of the main conclusions to be drawn, therefore, is that notwithstanding the existence of a procurement policy with a local content, this policy was unable to promote the production of high-value-added, R&D-intensive equipment. The predominance of multinational EPC contractors in Petrobras's platform projects also evidences both the almost complete absence of domestic shipbuilders and EPC contractors, and the technological and entrepreneurial fragility of the Brazilian industrial system.

As for the EPC contractors, subcontracted firms supplied mainly products that involved routine activities, so that learning by doing was the most recurrent kind of learning identified. One of the obstacles to innovative learning is the low technological capacity of the BOSI. Their limited experience in basic product design forces them to license foreign technology or use designs developed by the customer (in this case Petrobras–CENPES) or

EPC contractors. Therefore, the main conclusion of the study is that, despite the uniqueness of integral construction of a semi-submersible platform in Brazil, Petrobras has adopted a low-innovative PPI strategy. In addition, the fact is that the top procurement priority for Petrobras, rather than promoting innovation learning among Brazilian suppliers, is increasing local content, which does not contribute to a change in these firms' technological dependency.

Accordingly, Petrobras's procurement policy nowadays repeats the import substitution strategy, by emphasizing local content requirements. Nevertheless, such a policy does not stimulate the technological capability of local suppliers, so that the technological dependence of these companies is not eliminated. The authors consider that Petrobras could contribute to an improvement in this situation, given its status as a technological cutting-edge company, with respect to the exploration and production of oil and gas in deep waters, and the massive investment it will have to make to extract oil profitably from the pre-salt layer far below the ocean bed, its new production frontier. However, Petrobras has so far shown reluctance to assume the role of promoting technological development among local suppliers because of the risks and costs involved in doing so.

In order to encourage Petrobras to adopt an innovation-oriented procurement policy, the Brazilian government must agree to bearing this extra cost. Thus the main thesis advocated by this chapter is that given the technology lag suffered by firms in developing countries, a state-owned enterprise cannot alone bear the burden of fostering innovation by firms via advanced PPI. Developing-country governments should make good use of their industrial and technological policy arsenal (such as long-term financing, tax exemption, R&D subsidies and so on) to help local firms participate in public sector procurement of items that constitute innovations, and acquire higher levels of technological capability through a continuous learning process.

NOTES

1. It is worth noting that this kind of instrument is frequently referred to as government technology procurement policy (Aschoff and Sofka, 2009) or public technology procurement policy (Edquist and Hommen, 1998). The

term used in this chapter, public procurement for innovation (PPI), is preferred as more suitable to characterize a policy that promotes technological innovation.
2. Offshore projects involve oil extraction from the seabed.
3. The oil supply industry is constituted by the set of firms that, in addition to providing inputs and equipment, outsource manpower and services to the oil companies in the upwind activities of the oil chain.
4. Petrobras is the largest Brazilian company and a fundamental instrument of the federal government energy, industrial and innovation policies. The target to attain oil self-sufficiency was not completely accomplished until recently, despite national oil covering more than 98 per cent of the country's oil needs in 2011 (EPE, 2012). The petrochemical arm of Petrobras (Petroquisa) was privatized at the beginning of the 1990s. However, the constitutional reform of the mid-1990s only opened the Brazilian oil and gas market to private firms, including multinationals. The company was kept under federal government control in spite of the sale of part of the company's shares in the international stock market.
5. Semi-submersible platforms are made up of a superstructure with one or more decks, supported by ballasted watertight pontoons located below the ocean surface (Petrobras, 2007).
6. EPC (engineering, procurement and contracting) contractors are engaged by oil companies to supply services comprising detail engineering, procurement (of materials, equipment and services) and construction, for example in offshore platform projects.
7. In some cases, the public sector purchases are items to be used by the population of the country. In this case we are talking about catalytic procurement.
8. For a discussion of success cases, see for example Gregersen (1992), Langlois and Mowery (1996), Fridlund (1997) and Llerena et al. (2000).
9. A federal mixed-capital corporation is a legal entity established by a specific act of Congress as an instrument of state action, with a private-law personality but subjected to certain special rules that derive from this government action of an auxiliary nature, and constituted as a corporation with the federal government or a semi-autonomous agency as majority owner of voting share and the rest in private hands (Mello, 2002).
10. At the end of 2002, during the end of Cardoso's government, came the publication of the auction notice of P-51. In February 2003, shortly after the inauguration of President Lula, Petrobras's management decided to suspend the bidding process for P-51, which was already under way, to include in the notice the requirement of minimum national content in various parts of the platform such as the hull.
11. Marlim Sul, discovered in 1987 through the RJS-382 well, is located about 120 km off the northern shore of Rio de Janeiro state, in water depths ranging from 800 to 2400 metres, and occupies an area of roughly 884.1 sq. km (Petrobras, 2008).
12. Petrobras selected the P-51's EPC contractors from the analysis of the technical and commercial proposals developed by the companies invited to participate in the bid for this oil platform.
13. Rolls-Royce is a United Kingdom-based multinational public holding company that, through its subsidiaries, designs, manufactures and

distributes power systems for customers in civil and defence aerospace, marine and energy markets.

14. Nuovo Pignone is a company belonging to the business arm of GE, called GE Oil & Gas. This company has headquarters in Italy and, in addition to being a manufacturer of equipment (such as compressors), is dedicated to the building of gas compression and energy generation for oil platforms.
15. The FSTP Consortium included the firms Keppel Fels and Technip. Singapore-based Keppel Fels is a global leader in shipbuilding and oil rig production, with a long track record as a Petrobras supplier. In Brazil it owns the KeppelFELS shipyard and manages the BrasFELS shipyard, located in Niterói and Angra dos Reis, respectively. Technip, a French firm, is a world leader in engineering, technology and project management for the oil and gas industry, with an engineering office in the city of Rio de Janeiro.
16. FEED Endorsement is the stage of the project in which EPC contractors are able to analyse the documentation received from Petrobras to see whether they have any technical questions about their responsibilities.
17. It is worth noting that the FSTP Consortium was also the main EPC contractor for P-52. Work on construction of this floating production unit began before the start of work on P-51, since P-52 was installed in the Roncador field, which contains lighter oil that is more profitable than the heavy oil in Marlim Sul, where P-51 was installed. The timetable for completion of the two projects overlapped so that for a time the Brasfels shipyard was working on both projects concurrently.
18. A dry dock is a narrow basin on the shore of the sea or a river that can be flooded to allow a load to be floated in, then drained to allow that load to come to rest on a dry platform. Dry docks are used for the construction, inspection, maintenance and repair of ships, boats and other watercraft.
19. A long steel arm projecting off the side of the platform, used to burn off (flare) waste gases.
20. IBGE classifies Brazilian firms by size based on the number of employees. According to this methodology, industrial firms are divided into four groups: micro (up to 19 employees), small (up to 99 employees), medium (up to 499 employees), and large (500 or more employees). The classification for service providers is slightly different: micro (up to 9 employees), small (up to 99 employees), medium (up to 499 employees), and large (500 or more employees).

REFERENCES

Alveal Contreras, C. (1994). A Petrobrás: os desbravadores e a construção de Brasil industrial. Rio de Janeiro: Relume Dumará: Anpocs.

Aschoff, B. and Sofka, W. (2009). Innovation on demand – can public procurement drive market success of innovations?, *Research Policy*, 38(8), 1235–1247.

Audet, D. (2002). The size of government procurement. *OECD Journal*

on Budgeting, 4(1), http://www.oecd.org/dataoecd/34/14/1845927.pdf, accessed 22 December 2008.
Barlow, J. (2000). Innovation and learning in complex offshore construction projects. *Research Policy*, 29(7), 937–989.
Bell, M. (1984). Learning and the accumulation of industrial technological capacity in developing countries. In Fransman, M. and King, K. (eds), *Technological Capability in the Third World*. London: Macmillan, pp. 187–210.
Bell, M. and Pavitt, K. (1993). Technological acculumation and industrial growth: contrasts between developed and developing countries. *Industrial and Corporate Change*, 2(2), 157–210.
Campbell, J. (1988). *Collapse of an Industry: Nuclear Power and the Contradictions of US Policy*. Ithaca, NY: Cornell University Press.
Cohen, L. and Noll, R. (1991). *The Technological Pork Barrel*. Washington, DC: Brookings Institutions.
Cook, P.L. (1985). The offshore supplies industry: fast, continuous and incremental change. In Sharp, M. (ed.), *Europe and the New Technologies: Six Case Studies in Innovation and Adjustment*. London: Francis Pinter, pp. 208–217.
Dalpé, R. (1994). Effects of government procurement on industrial innovation. *Technology in Society*, 16(1), 65–83.
Dantas, A.T. (2004). Capacitação tecnológica de fornecedores em redes de firmas: o caso da indústria do petróleo offshore no Brasil. PhD dissertation, Universidade Federal do Rio de Janeiro, Rio de Janeiro.
Dosi, G. (1988). The nature of the innovative process. In Dosi, G., Freeman, C., Nelson, R., Silverberg, G. and Soete, L. (eds), *Technical Change and Economic Theory*, London: Pinter Publisher, pp. 221–238.
Edler, J. and Georghiou, L. (2007). Public procurement and innovation – resurrecting the demand side. *Research Policy*, 36(7), 949–963.
Edler, J., Ruhland, S., Hafner, S., Rigby, J., Georghiou, L., Hommen, L., Rolfstam, M., Edquist, C., Tsipouri, L. and Papadakou, M. (2005). Innovation and public procurement. Review of issues at stake: study for European Commission. No. ENTR/03/24. Final Report. Fraunhofer ISI, Karlsruhe.
Edquist, C. and Hommen, L. (1998). *Government Technology Procurement and Innovation Theory*. Linköping: Linköping University.
Edquist, C. and Hommen, L. (2000). Public technology procurement and innovation theory. In Edquist, C., Hommen, L. and Tsipouri, L. (eds), *Public Technology Procurement and Innovation*. Boston, MA, USA; Dordrech, Germany; London, UK: Kluwer Academic Publishers, pp. 5–70.
Edquist, C. and Zabala-Iturriagagoitia, J.M. (2012). Public procurement for innovation as mission-oriented innovation policy. *Research Policy*, 41, 1757–1769.

Empresa de Pesquisa Energética (EPE) (2012). Brazilian energy balance – year 2011. Rio de Janeiro: EPE.
Figueiredo, P.N. (2002). Does technological learning pay off? Inter-firm differences in technological capability-accumulation paths and operational performance improvement. *Research Policy*, 31(1), 73–94.
Figueiredo, P.N. (2004). Aprendizagem tecnológica e inovação industrial em economias emergentes: uma breve contribuição para o desenho e implementação de estudos empíricos e estratégias no Brasil. *Revista Brasileira de Inovação*, 3(2), 323–362.
Figueiredo, P.N. (2006). Capacidade tecnológica e inovação em organizações de serviços intensivos em conhecimento: evidências de institutos de pesquisa em Tecnologias da Informação e Comunicação (TICs) no Brasil. *Revista Brasileira de Inovação*, 5(2), 403–454.
Fransman, M. (1984). Technological capability in the third world: an overview and introduction to some of the issues raised in this book. In Fransman, M. and King, K. (eds), *Technological Capacity in the Third World*. London: Macmillan, pp. 3–30.
Freitas, A.G. (1999). Processo de Aprendizagem da petrobras: Programa de Capacitação Tecnológica em Sistemas de Produção Offshore. Campinas: Unicamp, PhD dissertation.
Fridlund, M. (1993). The development pair as a link between systems growth and industrial innovation: cooperation between the Swedish State Power Board and the ASEA Company. Stockholm: Royal Institute of Technology, Department of History of Science and Technology.
Fridlund, M. (1997). Switching relations: the government development procurement of a Swedish computerized electronic telephone switching technology. Report of research project funded by the Targeted Socio-Economic Research (TSER) programme of the European Commission (DG XII) under the Fourth Framework Program. Linköping University Sweden: European Commission Systems of Innovation Research Programme (SIRP).
Furtado, A. (1994). *Le Système d'Innovation Français dans l'Industrie Pétrolière*. Paris: CNRS.
Furtado, A.T. (1999). Petrobrás: Une reussite dans l'offshore profond (Petrobras: a success in the deep offshore). *Revue de l'Energie*, 503(1), 35–41.
Furtado, A.T., Marzani, B.S. and Pereira, N.M. (2003). Política de compras da indústria do petróleo e gás natural e a capacitação dos fornecedores no Brasil: o mercado de equipamentos para o desenvolvimento de campos marítimos. Rio de Janeiro: Tendências Tecnológicas – Projeto CT-Petro.
Georghiou, L., Amanatidou, E., Belitz, H., Cruz, L., Edler, J., Edquist, C., Granstrand, O., Guinet, J., Leprince, E., Orsenigo, L., Rigby, J., Romaneinen, J., Stampfer, M. and Van Den Biesen, J. (2003). Raising

EU R&D intensity: improving the effectiveness of public support mechanisms for private sector research and development: direct measures. Brussels: European Commission (EUR 20716).

Geroski, P.A. (1990). Procurement policy as a tool of industrial policy. *International Review of Applied Economics*, 4(2), 182–198.

Gregersen, B. (1992). The public sector as a pacer in national systems of innovation. In Lundvall, B. (ed.), *National Systems of Innovation: Towards a Theory*. London: Pinter, pp. 129–144.

Hoekman, B. and Mavroidis, P. (1995). The WTO's Agreement on Government Procurement. Washington, DC: World Bank.

Katz, J. (1974). Importación de tecnología, aprendizaje local y industrialización dependiente. México: Fondo de Cultura Económica.

Katz, J. (1981). Importación de tecnología e desarrollo dependiente. *Serie de Lecturas*, 38(2), 193–213.

Katz J. (1987). Domestic technology generation in LDCs: a review of research findings. In J. Katz (ed.), *Technology Generation in Latin American Manufacturing Industries*. London: St Martin's Press, pp. 13–37.

Lall, S. (1978). Transnationals, domestic enterprises, and industrial structure in host LDCs: a survey. *Oxford Economic Papers*, 30(2), 217–248.

Lall, S. (1980). Developing countries as exporters of industrial technology. *Research Policy*, 9(1), 24–52.

Lall S. (1982). Technological learning in the third world: some implications of technology export. In Stewart, F. and James, F. (eds), *The Economics of New Technology in Developing Countries*. London: Frances Printer, pp. 157–179.

Lall, S. (1992). Technological capabilities and industrialization. *World Development*, 20(2), 165–182.

Lall, S. (1994). The East Asian miracle: does the bell toll for industrial strategy? *World Development – Special Issue on the East Asian Miracle*, 22(4), 645–654.

Langlois, R.N. and Mowery, D.C. (1996). The federal government role in the development of the American software industry: an assessment. In Mowery, D.C. (ed.), *The International Computer Software Industry: A Comparative Study of Industrial Evolution and Structure*. New York: Oxford University Press, pp. 53–85.

Llerena, P., Matt, M. and Trenti, S. (2000). Public technology procurement: the case of digital switching systems in France. In Edquist, C., Hommen, L. and Tsipouri, L. (eds), *Public Technology Procurement and Innovation*, Vol. 16, Boston, MA: Kluwer Academic Publishers, pp. 197–216.

Macedo e Silva, A.C. (1985). PETROBRÁS: a consolidação do monoólio estatal e a empresa privada (1953–1964). Campinas: UNICAMP, Masters dissertation.

Mello, C.A.B. (2002). Sociedades Mistas, Empresas Públicas e o Regime de Direito Público. *Fórum Administrativo Direito Público, São Paulo*, 1(17), 871–875.

Nelson, Richard and Winter, Sidney (1977). In search of useful theory of innovation. *Research Policy*, 6(1), 36–76.

Oliveira, A. De and Rocha, F. (2008). Conclusões e recomendações de política. Estudo da Competitividade da Indústria Brasileiras de Bens e Serviços do Setor de P&G, http://www.prominp.com.br/pagina dinamica.asp?grupo=245, accessed 25 April 2009.

Petrobras (2007). Tipos de plataformas. Rio de Janeiro. http://www2.petrobras.com.br/portal/frame.asp?pagina=/Petrobras/portugues/plataforma/pla_plataforma_operacao.htm&lang=pt&area=apetro bras, accessed 7 April 2007.

Petrobras (2008). Campo petrolífero de Marlim Sul – Desenvolvimento da produção de petróleo e gás natural. Rio de Janeiro. http://www2.petrobras.com.br/portal/frame_ri.asp?pagina=/ri/port/Destaques Operacionais/ExploracaoProducao/ExploracaoProducao.asp&lang =pt&area=ri, accessed 26 June 2008.

Reis, O. (2004). Estratégia de compras locais para a construção da P-51, http://www.onip.org.br/arquivos/Estrategia%20Compras%20 Locais%20P-51%20-%20Rev%20C.pdf, accessed 30 May 2007.

Ribeiro, C.G. (2005). A Política de Compras de Entidades Públicas como Instrumento de Capacitação Tecnológica: o Caso da Petrobras. Master's dissertation, State University of Campinas, Campinas.

Ribeiro, C.G. (2009). Compras Governamentais e Aprendizagem Tecnológica: Uma Análise da Política de Compras da Petrobras para seus Empreendimentos Offshore. PhD dissertation. Universidade Estadual de Campinas, Campinas.

Rolfstam, M. (2005). Public technology procurement as a demand-side innovation policy instrument. DRUID Academy Winter 2005 PhD Conference, Lund, January.

Ruas, J.A.G. (2010). Transformações na concorrência, estratégia da Petrobras e desempenho dos grandes fornecedores de equipamentos subsea no Brasil. In: Poder de compra da Petrobras: impactos econômicos nos seus fornecedores. Convênio Petrobras/IPEA.

Surrey, J. (1987). Petroleum in Brazil – the strategic role of a national company. *Energy Policy*, 15(1), 7–21.

Villela, A.V. (1984). Empresas do Governo como Instrumento de Política Econômica: os Sistemas Siderbrás, Eletrobrás e Telebrás. Relatório de Pesquisa, nº 47, IPEA/INPES, Rio de Janeiro.

Weiss, L. and Thurbon, E. (2006). The business of buying American: public procurement as trade strategy in the USA. *Review of International Political Economy*, 13(5), 701–724.

11. Conclusions: lessons, limitations and way forward

Jakob Edler, Charles Edquist, Nicholas S. Vonortas and Jon Mikel Zabala-Iturriagagoitia

As stated in the Introduction, this book aims to provide a well-rounded understanding of the key determinants in implementing effective public procurement initiatives to achieve innovative outputs. The book provides both case studies and conceptual contributions that help to extend the frontiers of our understanding in areas where there are still significant gaps. The contributions are deliberately broad and diverse to show the range of issues that still need better understanding and reflection regarding public procurement for innovation (PPI). Out of this range of issues we may extract a set of key messages and implications for further research, procurement practice and innovation policy.

The contributions remind us of an important distinction between, on the one hand, the policy of using public procurement to spur innovation as part of demand-side policies, and on the other hand, the public procurement practice that aims to solve a specific societal problem or improve a certain public service and in doing so asks for (and commits to buy) something new. The two are often linked, but PPI as a policy instrument cannot be thought of without the procurement practice on the ground. PPI as a demand-side innovation policy instrument is a systematic attempt by authoritative public bodies to mobilize the purchasing power of the state (national, regional) for innovation policy goals. PPI as practice on the ground has a different logic; the innovation is a necessary means for achieving some other goals not necessarily related to innovation policy.

This distinction has certain implications: (1) isolated PPI practice on the ground cannot be assessed against the innovation policy criteria of PPI as a policy instrument; (2) policy learning needs to be derived from the former to the latter; and (3) the remit of PPI should be broadly understood. PPI can constitute a type of policy in its own right, with its own frameworks and innovation-related goals and even its own specialized agencies. However, PPI can also be understood as a policy instrument that seeks to uplift the capabilities within procuring bodies, and improve the framework conditions to enable the general public procurement practice to ask for and buy more innovations. The book contains illuminating examples of both cases with important overall policy lessons.

LESSONS FROM PPI AS AN INNOVATION POLICY TOOL

Four of the chapters clearly deal with PPI as a policy tool. In the country case of the USA, there is a Federal Framework that in some areas is designed to use public procurement as a policy instrument to spur innovation. As Vonortas has shown in Chapter 6, a similar basic framework across different policy areas at the federal level has very different results. For example, the stated PPI policy in the case of the environment works better than in other areas. The key lessons that can be derived from the US example are, among others, the idea of innovation generation being framed as an integral element to pursue the 'social cause' of environmental protection, and the PPI policy being designed through a bundle of measures, from regulation to 'policing' of practice, as well as guidance and information.

Similarly, the Chinese example of the procurement of electrical vehicles by Li et al. (Chapter 7) shows how PPI is used to improve the further development of those vehicles and accelerate their dissemination. These are classical innovation policy aims, and the means used are public procurement mechanisms. However, this example is reminiscent of the inbuilt tensions when public procurement is designed as an innovation policy tool first and foremost geared towards the economic effect of the innovation, rather than the societal and public service effect. In these cases, policy makers are seeking maximum innovation return within their own constituencies at the expense of maximum benefit for public service.

The chapter shows that while individual procurement processes within the scheme have worked, and commercialization of e-vehicles has been fostered, the overall economic benefit has been far more limited than expected. The reasons are important general lessons for PPI policy: first, protectionism by the regional buying bodies suppressed the broader competition that would lead to better solutions across the country. Second, as the national level did not have the leeway to control implementation, the attempt to govern through quantitative targets failed. This is a major challenge of PPI as policy; it is often designed at a higher, political level, while the implementation happens at lower levels, where interests are more local, capabilities more limited and additional support and incentives, such as co-funding and educating of procuring bodies (the demand side), are essential.

The forward commitment procurement (FCP) scheme introduced by van Meerveld et al. in Chapter 5 offers a similar lesson to the Chinese example. FCP has been designed as tool offering procuring bodies a means to reduce risk and improve interaction and expectation management. Here, we have a very simple but powerful first lesson to learn: it is possible to design mechanisms that tackle some of the core, fundamental challenges of procurement practice, that is, the management of risk and expectations, giving suppliers a firm and reliable goal to work towards, as there is a purchase at the end of a first-stage development process. However, a second lesson, linking to the Chinese example, is hidden: the instrument has not really been rolled out; we observe again a centrally designed instrument that is geared towards innovation, and the lack of uptake of this instrument across procuring bodies. PPI policy that should work across constituencies, not only for dedicated individual ministries or organizations, needs to ensure that the broader conditions are met for public bodies to take advantage of the instrument that is designed for them at a central level.

The case of Petrobas (Ribeiro and Furtado, Chapter 10) is a final example of PPI as a deliberate innovation policy. It offers a similar lesson, but with supply-side implications. Here the procurement power and a specific design of the procurement process of a state-owned company are used in order to uplift the technological competencies of incumbent firms. Again, this is a core innovation policy goal, which is linked to a whole range of further policy goals. The initiative largely failed in this innovation goal, as the incumbent local firms in supply chains did not have the learning capacity in place. Mobilizing firms into innovation procurement initiatives through the supply chain, and hoping for an innovation uplift, did not work. Firms need a basic level of innovation capacity in the first place to benefit from innovation procurement. The policy lesson here is clear: if PPI is to uplift the innovation capacities of firms, this uplift itself needs to be prepared and supported with appropriate supply side measures.

POLICY LESSONS FROM PPI PRACTICE TO ACHIEVE ORGANIZATIONAL AND POLICY IMPROVEMENT

Two examples are devoid of deliberate innovation policy initiatives, but deliver important lessons for PPI policy. The e-government case in Greece (Caloghirou et al., Chapter 8) and the recycled paper case (Yeow et al., Chapter 9) are first and foremost initiatives to improve public service provision. Part of the delivery of this improvement is the procurement of new goods and/or services. In both cases, the basis for the procurement process is not a classical innovation policy rationale, and the innovation as such is not the success metric here: the efficient and effective provision of the service is. Of course, trying to understand what works and what does not work when procuring innovation, we look at the innovation effects (or the lack of them) and try to derive important lessons for a more general support of public procurement of innovation.

The e-government case offers us an insight into a mechanism which seeks to develop an innovative solution at one entry point and then roll out this solution more broadly. The initial procurement is the test bed or a systemic change. The case reconfirms the importance of champions, of lasting relationships between suppliers and demanders. The case demonstrates how technological lock-in can be avoided when buying highly innovative systems through open source and open standards. However, it also illuminates the limits of central initiatives that are to be rolled out, as the lack of capabilities and user engagement, as well as resistance to intra-organizational change, are limiting factors at lower levels, aggravated by poor supply chain involvement and management at local level.

The closed loop recycled paper case teaches us the importance of problem pressure for PPI practices to be initiated and to actually work, up to a point where the risk of non-innovation is perceived to be higher than the risk of failing with an innovation. While the case also focuses on the importance of leadership, it reminds us of the tension of lasting relationships between producers and the buying organizations, both enabling (joint learning) and stifling innovation (through lock-in, lack of competition).

POLICY LESSONS FROM CONTRIBUTIONS FOCUSED ON BARRIERS IN PROCUREMENT PRACTICE MORE BROADLY

Three contributions in the book do not take concrete procurement cases or countries into focus, but specific challenges. Those challenges must be at the heart of PPI policy. In their survey of suppliers, Edler et al. (Chapter 2) show that, in principle, there are practices and processes out there that deliver innovation across the public sector, but public bodies are not able and/or willing to apply them. A core challenge that is at the root of many of those organizational practices is risk aversion. In this sense, Edler et al. (Chapter 4) make a first attempt to understand risk conceptually and call for a differentiation of risk for different actors and in different stages of a procurement process. Only if we understand those risks in detail can we devise policy strategies within PPI policies that manage risk in a productive way. In turn, Valovirta (Chapter 3) makes a first step in developing concrete management requirements based on the understanding of multiple challenges for procuring organizations, which we may read as a manual for procuring organizations.

These, and several of the application-relevant observations scattered across the chapters, treating specific procurement cases or country approaches, tie in quite well with the challenges stressed in the introductory chapter section on the operationalization of public procurement for innovation. Three of those challenges stand out as needing attention:

1. Translation of the identified societal goal of public procurement into functional requirements. Long experience points to the utility of 'functional procurement': the procuring agency formulates requirements in functional terms rather than in product terms, thus allowing the prospective suppliers to use their creativity and innovativeness to meet the desired

functionality. A prerequisite for success is the early engagement of stakeholders and expert advisory bodies. Important threats here originate in 'special interests', including the interests of politically influential organizations married to particular technological solutions (trajectories) and/or the interests of influential social groups.
2. Translation of the functional specification into technical requirements. While it is prudent to let prospective bidders translate the desired functionalities into technical specifications, such practice is antithetical to the incentives of purchasing managers bent on minimizing risk exposure. Risk reduction can be achieved by either procuring off-the-shelf products – thus also achieving short-term cost reduction objectives – or by detailing technical specifications. None of these is harmonious with innovation. In order to incentivize public employees to take on more risk, elected officials must offer protection by assuming the responsibility associated with spending taxpayers' money.
3. Delivery process. This stage of procurement encompasses product development, production and final delivery to the purchasing agency. While time and cost overruns are not unusual in public procurement, they become more acute in PPI due to the uncertainties of the targeted programmes in radical, game-changing technologies. The inevitable interference of the political system to accommodate the evolving, and occasionally contrasting, interests of stakeholders also exacerbates the problem.

We do hope that the concepts, insights and lessons presented in this book are instrumental in supporting policy makers and procurement practitioners in their effort to better use the public purse to initiate and support innovation for economic growth and, more broadly, societal welfare. This book fills gaps, but also opens new avenues. We can finish with some bold but simple messages for policy and academia:

1. PPI as an innovation policy instrument must be based on a good understanding of the challenges down at the organizational level and the different interests at different policy levels. While large innovation procurement initiatives may make a difference, the real difference is made if the whole innovation system is uplifted, if all levels understand the

issues required, and if the public sector puts emphasis on and opens debate with multiple stakeholders in order to deal with them. Learning from PPI practice, as shown in this book, is a first step in this direction.

2. PPI as innovation policy must be based on a good understanding of the need for absorptive and innovative capacity, and thus develop systemic approaches. PPI must be embedded into a range of policy instruments, from both demand and supply sides, that provide the requirements for it to improve its effectiveness in mitigating social demands or improving agency needs.
3. PPI policy must be able to ensure that the challenges identified at the practice level can be overcome by actors at all levels: by those implementing procurement initiatives, by those supplying them and by those benefiting from them. Tackling societal challenges in isolated large-scale PPI initiatives will not deliver the breadth needed. Therefore, PPI must be understood not as an ensemble of grand initiatives, but as a systematic roll-out of schemes, instruments and framework conditions that overcome those challenges across the system. This would ensure that PPI practice in general turns towards an (implicit) tool for PPI and innovation policy.

The academic interest in PPI policy will continue, and must continue. While the innovation policy community certainly has knowledge and experience of policy instruments and challenges, more knowledge is needed about the organizational conditions that make some organizations less vulnerable than others. This book has provided a clear set of managerial implications, but a long bridge still needs to be crossed if the efficiency and the effectiveness of the public sector are to be further improved.

The interest in demand-side policy measures in general has gained momentum in the last few years, and so have particular instruments such as PPI. However, the evaluation of demand-side innovation policies is still in its infancy. Thus, contributions from both the policy evaluation community and the innovation policy community are needed in order to be able to develop comprehensive frameworks and methodologies, and to define the required indicators that may feed them. Only in this way will it be possible to achieve a co-evolution between the development of the innovation policy community and its related evaluation practice.

Index

acceleration speed of car, high, Shenzhen 196
Accessible and Personalised Local Authority Website System (APLAWS) 220–21
acquisition guidance, agency-level 153
acquisition officers in federal agencies,
  innovative procurement practices 159
  White House Office of Federal Procurement Policy (OFPP)
acquisition rules, FAR policy (US) 152
Acquisitions Offices 153
adaptive procurement capability 71–2
ADS-B *see* Automatic Dependent Surveillance-Broadcast
advanced innovative learning results 271, 279
aerospace industries 163–4
Aho Group Report, 2006 1–2
air traffic control, satellite-based system 163–4
Aker company, Norway 276
American Community Survey 163
American Recovery and Reinvestment Act, 2009 157–8
Audit Commission 41
Automatic Dependent Surveillance-Broadcast (ADS-B) program
  aircraft identification 163–8
  increased airspace capacity and efficiency 167
  services contract 164–5
award criteria 121
  carbon footprint, total cost of ownership 134
awareness-raising campaigns 104
AWIPS, data gathering by radar 173
barriers, artificial creation for new entrants 79
barriers, significance of 55
barriers to innovation procurement
  too much weight to price rather than quality 54
base registries, sources of information 213
bat swarms, mobile storm-chasing radars 173
battery configuration 192
bed washing innovation
  Erasmus University Medical Centre (Erasmus MC), the Netherlands 124
Beijing
  target of 5000 NEVs by end of 2015 185
biodegradable censors, NOAA 169
BOSI *see* Brazilian Oil Supply Industry
boundary layer turbulence, Hydro-Radar 173
bounded rationality
  incomplete information conditions 122
  proposition of TCE 122
  rational decision-making 122
BrasFELS shipyard, hull fabrication 282
Brazil, Petrobras company, case study data 264
Brazilian oil and gas supply, P.51 construction 265–6
Brazilian Oil Supply Industry (BOSI) 264
  innovative technological learning 273
  technological gap information 288–9
Brazilian Prime Vendor 288

Brazilian state companies 264
Electrobras, Siderbras, Telebras 265
Brazilian state intervention in economic affairs 273
Broadcast Services Ground System (BSGS) 165
building capability
public procurement for innovation (PPI) 65–83
Business, Innovation and Skills (BIS), Department 110
business process management system (BPMS) 222
buyer-supplier relationships 243
Buying Solutions 41
BYD pure electric buses (K9) 199
BYD pure electric taxis (E6) 199

cable-controlled underwater recovery vehicle (CURV) 1960s
ROV technology 170
Carbon Compacts 42
carbon emissions reduction, China 182
carbon footprint as award criteria 126–7
carbon reductions 127, 130
case studies innovation drivers 130
case studies overview 125, 130
Catalogues of Recommended Vehicle Models for NEV Demonstration Program, China 186–7
Census Bureau 163
conducting of US Census, 1980 problems 161–2
TIGER System 161–2
central government, innovation criteria in tenders 53
Central Union of Greek Municipalities (KEDE), 2007 209–10, 218–30
stages of procurement process 219–22
China
macro, national level of policy-making 187
medo, regional level of policy articulation 187
micro level of policy implementation 187
China Putian group 200
China Southern Power Grid Co. Ltd
research on existing standards 200
Chinese Law on Government Procurement 180
Chinese new energy vehicles programme 300–301
case studies 20
PPI 179–205
Cimarron Doppler radar 173
citizen expectations of service quality 66
city-level public procurement activities, China 201
clean air, water, natural resources concerns
in research and innovation 154
Clean Air and Clean Water Acts 154
climate change 70, 111
closed-loop paper model
case study data 244– 53
delivered by Banner 251
procurement process 249–53
production at paper mill 251
with zero waste 251
cloud computing system creation 159
Cockpit Based Merging and Spacing 167
Code of Federal Regulations 154
collaboration in PPI 9
commercialization of innovations 3
commercialization of NEV technology, China 182
communication and interaction in China 201–3
Community Support Framework 215
Company A, China, hybrid coaches 190–92
competition, full and open, FAR policy (US) 151
competitive dialogue 52–3, 121
competitive processes 148
complex performance procurement 66–7
Computer Aided Telephone Interviewing (CATI) 42
computer systems
petaflop capabiity 171–2
teraflops capability 171

confidentiality and investment, suppliers' concerns 134
conflict between policy objectives 38
conflicting goals of principals and agents 122
construction and assembly of metal structures 284
construction of semi-submersible platform in Brazil
  low-innovative PPI strategy of Petrobras 292
Continuous Descent Arrival (CDA) 167
contract, size and duration of 56
contract awarding 16, 82, 229
contracting, in federal public procurement 176
contracting out service production 65
Contracts and Acquisitions Management group
  of Department of Education 158
contractual arrangements, complex 75
copier paper, low-risk item 244–5
copier paper recycling, for UK government 235
Corporate Register of Goods and Services Suppliers 288
corporate social responsibility 238
cost efficiency, innate principle, FAR policy (US) 151–2
cost-efficiency in US 174–5
cost estimating 101
cost issue as barrier 54
cost rationalization 15, 148–9
cost sharing 104
cross-border service 228
cross-case analysis issues, Jinan and Shenzhen 202
cultural differences, contextual risk 91
curriculum development 82
customer information need 114
customer risks in PPI 112–13
customers, private and public 56

damage to environment, minimization 238
dashboard redesign of car, Shenzhen 196
data exchange and management layer
  secure exchange, security requirements 213–14
Decennial Census, USA 163
decision-making in PPI 88
decision-making power, procurement manager, US 153
deck mating operation, P-51 283
deep-sea rescue operation
  cable-controlled underwater recovery vehicle (CURV), 1960s 170
defence expenditure, high in Greece 215
defence sector 11
deforestation, growing impact 236
demand aggregation, informal approach 136
demand and supply, risk sources 91
demand-based innovation policies 17–18, 35
  PPI as crucial instrument 228
demand-side
  activities, public procurement for innovation 1–2, 4–5
  market risks 100
  measures 18, 38–9
  policies 88
  policy instruments 29–31
demographic structure of population, changing 70
Department for Business, Innovation and Skills (BIS)
  FCP method steps 117
Department of Defense, US 157
Department of Education 154
Department of Energy Acquisition Regulations (DEAR)
  incorporation of FAR rules 155
Department of Energy (DoE), US 154–5
Department of Health 130
Department of Health and Human Services (DHHS) 154
  public funder of research, US 157
Department of Interior (DoI). public infrastructure 154, 157
design uniqueness, challenge of 75

developing countries
learning process 264
stages of foreign technology learning 269
developmental procurement capability 71–2
diffusion process preconditions, essential function 75
Digital Agenda for Europe 231
European Commission, 2010 212
digital database 162
digitalization of government activities 65–6
digital sign processing hearing aids in NHS 242
digital skills lack, Greek employees 218
digital spatial database, map data 162
Directgov 41
now Gov.UK website 61
disadvantaged groups, assistance to 148
domestic industry development 266
domestic shipbuilders, absence of
in Petrobras's platform projects 291
Doppler radar 172–3
Dual Independent Map Encoding (DIME)
geographic information system for spatial data 161
dual polarization radars 172–3
duplicate production, China 20

e-auction, e-tendering 53–4
economic development of disadvantaged groups 69
economic impacts of PPI 66
economic judgement for assessment 16
economic zones, special, Deng Xiaoping 204
ecosystems, residential 101
Efficiency and Reform Group (ERG) 41
Efficiency Review, Sir Philip Green (2010) 240
e-government and public services 210–12
Europe 211–15
Greek case 217–18, 302–3
e-government interoperability, significant benefits 214
electric cars, 'government renting', China 195
electric coach model, BYD 198
electricity generators 280
electric vehicle (EV) manufacturers in Jinan
low-speed vehicles for rural areas 189
employment and promotion of small businesses 69
end-users' involvement, limited, LGAF project 8, 224
energy 155–6
and water supply 111
conservation 148
efficiency, federal procurement (US) 151
efficient norms for buildings 70
independence and development 155
plants, alternative 101
Energy Policy Act 154
Energy-saving and New Energy Vehicles Demonstration, Promotion and Application Program (NEV program) 184
energy sector 11
energy technologies, alternative, US 155
engineering as main evaluation method
FAR policy (US) 152
environment 69, 148, 153–5
environment-friendliness, federal procurement (US) 151
Environmental Innovations Advisory Group (EIAG) 113–15
forward commitment procurement (FCP) 116
'mountain of risk' design 115
environmentally sensitive areas
transport, energy, food, paper 238
environmental modelling program (EMP). NOAA 171
Environmental Protection Agency (EPA) 153–4
Office of Acquisition Management 154

EPA Acquisition Regulations (EPAAR) 154
EPC contractors learning experience in P-51 project 281
EPC (engineering, procurement, contracting) 291, 293
Erasmus University Medical Centre (Erasmus MC) the Netherlands 125
  bed cleaning facility procurement strategy 126–7
  bed washing innovation 124–7
    market engagement 125–6
  competitive dialogue and procurement procedure 134
  initial ideas rather than detailed proposals 134
  market meeting day 126, 133
  market sounding steps 132
e-tenders 53
EU Cohesion Policy Fund, pilot project targets 219–20
EU-national projects, co-funded 216
European Commission reports
  e-government Benchmark Report, 2013 212
  Risk Management in the Procurement of Innovation 92
European Directives on procurement processes 15–16
  proposed changes 14
  water, energy, transport, postal services sectors 14
European Interoperability Framework 221, 231
  conceptual model for public services 213
European Investment Bank 102
European Parliament, 2014
  new directives on public procurement 9
European single market 214
European Union and PPI 88
evidence data for closed-loop model case
  in-depth interviews with stakeholders 244–5
  interviews with key individuals 244
  site visits 244
expenditure of UK public bodies, on procurement 237
export opportunities 37

FAA *see* Federal Aviation Administration 164–5
FAA Life cycle management process for ADS-B 168
failure causes and consequences 97–8
failure risk 111–12
failure tolerance pluralist political systems 15
FCP *see* forward commitment procurement (FCP)
Federal Acquisition Regulation (FAR) 153, 174
  acquisition mechanisms, non-defense federal agencies 153–9
  cost-effectiveness, open competition requirements 153
  in US 14, 20
federal agencies
  Departments of Education, Energy, Health and Human Services 149
  differences in federal public procurement 176
  Environment Protection Agency, Department of Interior 149
Federal Aviation Administration (FAA)
  Automatic Dependent-Broadcast (ADS-B) program 163–8
federal Department of Education (DoEd) 158
federal funding for education 158
federal government of US
  world's largest purchaser of goods and services 150
federal procurement 149–51
federal public procurement in US 174
FEED and detail engineering 280
financial crises 111
financial institutions, public, German KfW 102
financial management tightening, discouragement of innovation 149
financial risks 93, 102

firm sizes
by area of government 45
by procurement category 45
fish and wildlife, Department of Interior (DoI) 157
flare bloom projects 283
flexible corrective action 104
flight information, important 166
focus groups for innovations, for need articulation 12
foreign technology licensing 286
forensic detail practices 58
formal bids by potential suppliers 15
forward commitment procurement (FCP) 19–20, 42, 114–21, 301
actions, assessment of individual 135
advance information of future needs 116
cost savings and competitive processes 151, 174
decrease of client's uncertainty 137
demonstration project 116
development of strategy 118
effect on perceived risks in PPI projects 110–40
engagement, identification phase 117–18
overall effect of application 137–8
pilot projects 110
procedural rules 150–51
process for healthcare organizations 139–40
sections of relevance to PPI 151–2
theoretical framework 123–4
fossil fuels, hydraulic fracturing (fracking) 176
framework agreements, professional services 53
framework contracts 52–3
FSTP Consortium, EPC contractor, Brazil 282–3, 294
full life-cycle costing 52
function definition 13
functional requirements
performance-based, outcome-based, value-driven 76–7
translation into technical specifications 14
General Services Administration (GSA)
and Federal Acquisition Regulation (FAR) 151
geographic database, US 162
geographic information system (GIS) technology 163
geographic support of census
geographic reference files, maps 161
Geography Division: Topologically Integrated Geographic Encoding and Reference System (TIGER) geographic database 162
German paper mill 258
Global Positioning System (GPS) satellite navigation 163–4
global socio-economic challenges 66
goods, socially desirable, subsidies for private buyers 204
government demand, and innovation 49
government efficiency, UK 41
government entities, procurement by
leading role in domestic industries 290
government expenditure in GDP 147
government initiative, UK, paper waste recovery 235–6
government procurement, China 180
government procurement for developing countries, typology 272
government procurement policy 266–7
and domestic industry 263
government's market power 74
GPS signals, ADS-B 166
Greece, political hostilities, unfavourable for trust 215
Greek firm without technical competence 229
Greek information technology industry 216–17
Greek local authorities, e-government services 21
green themes demonstration 52, 201
ground radar and ADS-B systems, comparison 168

Ground Radar-Based System
comparison of cost with ADS-B 166–7

Health and Human Services Acquisition Regulation (HHSAR) 150–51, 157
healthcare requirements, commonality across Europe 136–7
high-performance computing (HPC)
NOAA environmental modeling program 171
HM Prison Service landfill of foam mattresses, annual 116
HM Revenue and Customs (HMRC)
closed-loop concept with Banner 250
commercial spending 243
copier paper, 100 % recycled content 252
data loss incident Poynter Review 249
data security, critical issue 253
internal audit 249
loss of discs in transit 243–4
low priority on information security 244
procurement of copier paper 248
procurement of waste management services 248
recycling and sustainable development 249
relationship with Banner, paper supplier 254–5
results of closed loop system 254
significant user of paper 245
HM Treasury 41, 242
Horizon 2020 programme, European Commission, 2013 237
horizontal alignment, between sector agencies 81
House of Lords Science and Technology Committee, 2011 42
hull block construction 280, 284
human resources lack for PPI practices in ICT
Greece 217–18
hybrid coaches 190–92
hybrid vehicles (NEVs) 182

ICT *see* information and communication technology
identification of (social or agency) needs 10–12
identification phase 131
for successful market engagement 138
impact of not innovating, risks 112
industrial boilers 288
industry inclusion in decision-making
in federal public procurement 176
information and communication technology (ICTs) 44, 210–12, 277
procurement in Greece 215–17, 230
information asymmetry 113, 122
infrastructure 102, 157–8
innovation
acceptance among users 80
activities, variety, service or process innovations 46
awareness, in federal public procurement 176
characteristics risks and remedies related to 100–101
commitment to 59
concept assessment of technical change in developing countries 264
criterion in procurement 158
different types of 46
environment, in energy sources, construction material, cleaning supplies 154
fostering, PPI efficiency 161, 267
generation process 69–70
ideas from supplier firms, processing 73
linear model's policy 3–4
piloting and testing 77–9
policy, PPI as effective tool 88
procurement 70–71
main barriers to 54–7
public demand for incentivization of industry 37
public procurement in the United States 147–78
public procurement, range of barriers 6, 57
public sector, effects 50

research, for energy sector, US 156
restraints and barriers 38
successful delivery 139
by suppliers, encouragement 241
systems, key activities 28
Innovation for Sustainability Programme 110
innovation-friendly public procurement 7–8, 54
innovation in the United States 147–78
innovation procurement plans (IPPs) 42
innovations
material goods or intangible services 3
new creations by firms 3
innovation source
public procurement for innovation (PPI) 47–8
innovations, radical, R&D funding 49
innovative learning, low, factors involved 279
Innovative Technology Adoption Programme (iTAPP) 42
Instituto Basileiro de Geografia e Estatistica (IBGE)
national statistics bureau, Brazil 280, 284
integration issues 65–6
intellectual property rights (IPR) 38, 75–6, 191
intellectual property rights (IPR) provisions, innovation fostering 52
interaction and exchange, developing trust 242
interaction with suppliers, capability 73
Interior Department Acquisition Regulations (IDAR) 157
intermediate-innovative learning, factors involved 279
intermediate learning
platform engineering design, equipment fabrication 286
International Cartographic Conference, 1984
nationwide geographic database, US 162
international collaboration projects 74
inter-regional competition, China 203–4
intra-organizational levels 81
invention of 'closed-loop paper' model 244
Investing in Innovation, grant program 158
investing incentive 56
investment needs 70–71
investment promotion in new products 116
investments in innovation by risk sharing 80, 113–14
ITT Corporation, contract award to 166

Jinan, China
procurement of hybrid coaches for National Games, 2009 188
Jinan city, avoidance of risky commitments 193–4
Jinan Public Transport Group 192

KEDE *see* Central Union of Greek Municipalities
KeppelFELS shipyard 282
knowledge inputs (research) 3
knowledge-intensive entrepreneurship opportunities 227

Laboratory of Industrial and Energy Economics (LIEE)
National Technical University of Athens (NTUA) 231
land management, Department of Interior (DoI) 157
large-scale projects 268
lawmaking
regulatory standards for environmental practices 154
LCB-Healthcare pilot projects 116, 124, 130
FCP method: engagement 131
Lead Market Initiative 88–9
lead markets
closed loop potential 252–3
creation of 49–50, 88

in NEVs, China 204–5
new products 74
research guarantees 104
in a sustainable innovation 237–8
learning between subcontractors 280–81
learning by design 271, 283
learning by doing 291
fabrication of process modules 283
learning experiences, typology of various
in developing countries 272
learning, interactive, between organizations
for innovations 12
learning mechanisms, formal and informal 82
learning modes by developing-country firms
low, intermediate, advanced 270–71
LED display system in bus, China 199–200
Leopoldo Américo Miguez de Mello Research and Development Center (CENPES) 275
LGAF *see* Local Government Application Framework
licensing of technologies 289
life cycle analysis (LCA) 238
life-cycle costing, as assessment criterion 14
linear view of innovation policy 5
litigation, extensive
in federal public procurement 175
Local Authority Websites National Project 220
Local Government Application Framework (LGAF)
character of 226–7
long-term potential 227–8
main actors 223
obstacles to success 224
path dependency as impediment 225
pilot project of Central Union of Greek Municipalities 21, 209–10, 218–26
platform 230
delivery to KEDE, 2011 222
used by Greek municipalities 228
redesign of 225–6
significant side-effect 227
local government contract clients 44
local service delivery 70
Lockheed Martin, ADS-B 164
Low Carbon Building (LCB), Healthcare Network 110
low-innovative learning by doing, activities 286
low-speed EVs, lead-acid batteries 189

main sources for innovation 47
management accountability, lack of
HM Revenue and Customs (HMRC) 244
managerial skills
for functional requirement development 77–9
manufacturer BYD, Shenzhen 194–5
market consolidation by PPI 148
market demand risks, assessments 101, 120
market engagement phase of FCP 131, 133–4
aligning procurement strategy 121
communication 119
consultation, personal 132–3
facilitating networking 120–21
identification of actions 119
market sounding 119–20
requirement refining 120
Rotherham NHS Foundation Trust (UK) 127–8
signposting demand 120
market escalation by PPI 148
market intelligence lack, policy need 60
market knowledge, lack of 59
market novelty, definition 39
market power of government buyers 73–4
market risks, demand and supply sides 93
market sounding prospectus (MSP)
direct emails, telephone calls, articles in trade journals 132–3

Rotherham NHS Foundation Trust (UK) 127
metal structures for the power generation 280
meteorology 172
military equipment procurement in Greece 215
military requirements 268
Ministry of Finance (MOF) 179
Ministry of Industry and Information Technology (MIIT), China 184–5
Ministry of Science and Technology (MOST) China 179, 182
mobile sensor platforms, NOAA 169
model-based analysis techniques, NOAA 169
monitoring of PPI 16–17
'mountain of risk' 115
multidisciplinary teams, use of 103
multiple sourcing, one contractor 16
multiple stakeholder objectives, possible contradictions 69

National Aeronautics and Space Administration (NASA) 160
National Airspace System Wide Acquisition Program Advance Notification to Industry, 2006
  quote on performance specification 165–6
National Audit Office (NAO) 238–9, 243–4
  on lack of data availability for procurement 242
  public spending monitoring 41
national defense and security areas, US 148
national demonstration programs, China
  lead market initiatives (LMI) 180
  type measures 180
National Development and Reform Commission (NDRC), China 184
National Games of China, 2009
  host city, Jinan, public transport 189
National Health Service (NHS) 41, 44–5
  difficulties in innovation 49
National Institute of Health (NIH)
  public funder of research, US 157, 160
National Medium- and Long-Term Program for Science and Technology Development (2006–2020) (MLP) 179
National Oceanic and Atmospheric Administration (NOAA)
  definition of activities 169
  high performance computing 169–73
  remotely operated vehicles 169–73
  scientific agency in US Department of Commerce 169
National Ocean Service (NOS)
  safety of ocean and coastal areas 170
  telecommunications and information technology 171–2
national policies for innovation procurement, China 181
national policy arenas 70
national problems, versus local problems 70
National Weather Service, NOAA reliance on radar 172
needs, future and unmet, lack of clarity 112
needs' 'translation' into functional requirements 12–13
Netherlands, academic hospital, pilot project 124
Netherlands Organization for Applied Scientific Research (TNO)
  project support, facilitation, know-how 130
networking, facilitating of, supply chain interaction 137
NEV *see* New Energy Vehicles
NEV-related innovation policies in China 183
  development in China 182–7
New Energy Vehicles (NEV)
  demonstration program 184–5
    procurement of 100 hybrid buses 189–92
  demonstration program in China 182–7
  'fever' 203

manufacturers in Shengzhen, BYD and Wuzhoulong 194
'green technology' 197
program design 186
standardization 200
procurement for Universiade 2011
process 196–201
new product development 137
new technology, dissemination of 88
Next Generation Air Transportation System (NextGen)
satellite-based system 163
NIH *see* National Institute of Health
NOAA *see* National Oceanic and Atmospheric Administration
computer capability 171–2
scientific productivity enhancement 171–2
non-departmental public bodies (NDPB) 41
Nordic countries, radical innovations in past 105
Nottingham University Hospitals NHS Trust (UK) 125
coal-fired boiler, end 128
energy solution, distributed 129–30
integrated ultra-low carbon energy solution 128–30
market engagement process 129, 133, 136
novel products, need for expected performance testing 77
nuclear power industry US, failure in PPI 267
Nuovo Pignone, sales office in Brazil 280

Obama, Barak, President of USA, major health reform 156
object recovery from ocean floor
cable-controlled underwater recovery vehicle (CURV) 1960s 170
ocean energy management
Department of Interior (DoI) 157
ocean-surface characteristics, Hydro-Radar 173
Office of Acquisition of Property Management (APM) 157
Office of Federal Procurement Policy (OFPP) 152
White House 175
Office of Government Commerce (OGC) 2004
'capturing innovation' 38
Office of Grants and Acquisition Policy and Accountability US 157
Office of Innovation and Improvement 158–9
Office of Management and Budget (OMB) at White House 151–2
Office of Chief Information Officer (IOCIO)
NOAA program and IT 171
Office of Government Commerce (OGC) 41
office paper supplies, collaboration in 246
Official Journal of the European Union (OJEU) 127
oil and gas industry, US, ROV technology 170–71
oil platforms
construction and assembly 279
engineering design 279
fabrication of equipment 279
functions 277
supply of services 279
oil refineries, Gabriel Passos, Alberto Pasqualini 273
oil savings rate of hybrid buses and cars 185, 192
oligopoly 79
open source software (OSS) 219–20
opportunism 122
proposition of transaction cost economics (TCE) 122
Organisation for Economic Co-operation and Development (OECD) 147
public procurement for innovation (PPI) 35–61
organizational benchmarking 82
organizational capability, definition 67
organizational risk 93, 103
organizational routines, repetitive 67–8
development of the platform 227–8

outcome specifications 52
in social services, professional services 54
outsourcing service production 65
overspecification of requirements, peril 159
in federal public procurement 175–6

P-51 new procurement policy of Petrobras 274
P-51 platform 274–90
P-51 semi submersible platform, 2009
bidding process for oil platform 276
contracts with EPC contractors 276–7
leadership UN Rio 275
offshore Rio de Janeiro state 274–5
P-51 subcontractors covered by field survey 282, 285
paper products buyer, and paper waste supplier 245
paper purchasing supply chain, typical 246
paper, recycled, lower price 252
paper recycling, economic and environmental benefits 236
paper shredding company 251
paper supply chain, from forest, or paper mill 245
paper waste, confidential, traceability and accountability 252
paper waste disposal 245, 247
paper waste generating, in UK public sector 236
partnership approach to public procurement
barriers to 78, 243
payment services by financial institutions 213
people, lack of well-trained
in federal public procurement 176
performance requirement for NEV in Universiade
set by Shenzhen government 197–8
personal data, loss of
HM Revenue and Customs (HMRC) 244
Petrobras, Corporate Register 284
state-owned enterprise, Brazil 280, 293
Brazilian oil and gas supply 264
deepwater oil production 291
front end engineering design (FEED) 275–6
loss of monopoly 273–4
'mixed-capital' corporation 273
outsourcing functions 274
procurement policy, case of P-51 273–4
platform engineering design 280
fabrication of equipment 279
policing by EPA 154
policy implementation, China 204
policy makers, elected, risk aversion of 81
policy targets, quantitative 203
political system interference 17
PPI *see* public procurement for innovation
PPOI, challenges and drivers 24043
precipitation monitoring by radar 172
pre-commercial procurements (PCP) 5, 9
pressure vehicles 280
private demand, insufficient 104
private finance initiative (PFI) contracts 248
private firms 11, 44
private research and development (R&D) leverage of 88
private sector companies, supply chains 116
private versus public customers, assessment of 57
process innovation 48, 147, 201
procurement cycle 93
procurement data
quality and availability in public sector 242
procurement effectiveness in fostering innovation 37
procurement expertise 241
procurement information, key barriers to 242
procurement in United States defense, security, social objectives 174

procurement modes and practices 51–4
influence on innovation 51
procurement of sustainable innovation
case of closed-loop paper 243–53, 257
procurement officers, emphasis on tools or purchase 160–61
procurement practices, use and meaning of 52–3
procurement procedures to support innovation 113
procurement process
core design principles 222–3
risks and remedies related to 103–5
procurement process of NEVs for National Games, 2009, China 190
procurement reform in US 175
procurement strategies 133–4, 241
procurer characteristics, assessment 55
procurer competence 241
procuring agency, 'functional procurement' 13
product development and final delivery to purchaser 17–18
product innovation 147
product-based services 227
professional services, increased R&D investments 49
profit impact of products 239
profit-sharing arrangements 51, 53
program initiatives, central-level NEV 201
project and process characteristics, risk management 100
protectionist safeguards 266
protectionism, aggravated, China 203
public bodies, importance as first customers 58
public buyers, less innovation-friendly 56–7
Public Contracts (Scotland) Regulations 2006 41
public demand as trigger for markets 37, 48
public expenditure on innovation in US 147
public health 111
in US science policy 156
public organizations 66, 71
public policy schemes 4, 60
public-private partnerships 248
public procurement for e-government in Greece 215–18
public procurement in environmental area
best examples of PPI on US 155
public procurement and innovation link 47–51
public procurement for e-government services
Greek local authorities 209–32
public procurement for innovation (PPI) 1, 36–40, 148, 237–8, 266–8, 551
catalytic or direct 8
consistency with other policies 69
definition, innovation promotion 6–7, 148
demand-side innovation policy instrument 299–300
in developing countries, case of Petrobras 263–94
different types 7–8
efficiency and effectiveness of money spent 69
in field of e-government 209
goal setting for 68–70
implementation stages 10
incremental and radical types 8–9
innovation policy, China, 2006 179
innovation policy instrument, needs 300, 304–6
interactive learning 9
intervention in China 14
low-innovative 272
major barriers 18
needful links with R&D 78
Petrobras, Brazilian enterprise case study 21–2
procedures, supplier unfamiliarity 134
processes, need for understanding 59
projects, from genuine unmet needs 140

public confidence in fairness of procedure 69
purchase of goods, services, by public agency 6–10
recycled paper for UK government 21
societal challenges 70
societal problem-solving 15
in the UK 40–42
in United States 160–73
Census Bureau 149
Department of Energy 149
energy sector 156
Federal Aviation Administration 149
matrix in US 150
National Oceanic and Atmospheric Administration 149
outside defense and national security 148–9
state and federal levels 150
public R&D funding 9–10
public sector
bureaucracies, risk aversion encouragement 112
contracts, important source of innovation 50
demand 11, 238
information systems, Greece, LGAF system 228
organizations 42–3, 111
potential, sustainability goals through innovation procurement 235
role, innovation generation process 46–7
public service improvement 66
public services reform UK 41
public spending categorization 39
public spending value for money 69
public technology procurement 2
public transport companies 191–2
purchasing portfolio (Kraljic), 1983 239–40
purchasing power, 159, 240
purchasing strategy 239–40
purpose of procurement 103–4

R&D design risk, reduction methods 101
R&D, developmental, underinvestment 78, 82, 268
R&D examples, prototype making, market research 61
R&D funding 12, 49
R&D phase 68, 104, 114
R&D support 147–8
radar object-detection system, definition 172
radical innovation concept 91–2, 269
Raytheon, ADS-B 164
recycled paper for UK government 257, 302–3
recycling plants, large 101
refining requirements, perceived actions 135–6
refurbishment programme
Rotherham NHS Foundation Trust (UK) 127
regional protectionism, China 20, 180, 193, 203
regional trade areas, European single market 74
regulatory complexity 38
relationship management
versus transactional contracting 78–9
remotely operated vehicles (ROV) 170
request for offers (RFO) 164
requirements in functional terms 13
requirements of procurement practices, writing of
success and failure 159
resource library, Environmental Protection Agency (EPA) 155
restrictions on item specifications FAR policy (US) 151
risk and innovation 19
risk averse behaviour, encouragement 112
risk aversion 55–6, 59, 87, 89, 303
risk conceptualizing in PPI 90–94
risk, contextual,
political instability, supply chain changes 89, 91
risk definitions 90
risk exposure 15

risk impact on FCP 121–4
risk in PPI 92–4
risk issue, importance of 111
risk management 19, 89, 123
basic functions and principles 94–100
reduction of innovation costs 94–5
risk management
conceptualization in PPI 94–105
in innovation procurement 66, 80
in PPI 87–140
lack 56
practices in public sector 87
tasks 99–100
risk map in PPI, sources of risk 95–6
risk of credible buyers and market delivery 131
risk sharing, through public funding 80
risk sharing among public sector 11
risks and uncertainties, perceived on case studies 131
risks at demonstration of product time
before commercial sales 114
risk-taking
different attitudes of principals and agents 122
Rolls-Royce, Brazil, sales office in Brazil 280
Rolls-Royce, UK-based company in Brazil 293
Rotherham NHS Foundation Trust (UK) 125
alignment process success 134
market consultation workshop 133
market engagement 127, 136
market sounding steps 132
pro-innovation procurement 128
ultra efficient lighting for future wards 127–8
ward solution 128
ROV *see* remotely operated vehicles

Sainsbury, Lord
UK Minister for Science and Innovation 115
satellite-based navigation 163–4
scale and complexity of PPI 101
scale, change of supply-side market risks 102
school performance, innovative techniques for 158
screening information request (SIR)
proposal submission 164
sedan models, BYD, plug-in hybrid car 198
selection criteria 121
semi-submersible platforms 293
service delivery, innovative products and services 36
service delivery missions
public health, safety, public transportation 68
service innovation 48
service platform on business processes 221
service-oriented architecture (SOA) 221, 226
service-product combinations 65
service sector activities 44
Shandong Government Procurement center 190–91
Shenzhen
charging piles for vehicles 200
target of 24000 by end of 2012 185
technologically advanced 194
Shenzhen Bureau of Power Supply 199
Shenzhen, Pearl River Delta NEVs, procurement for Universiade 2011 194–201
Shenzhen pilot project, 2008
hybrid buses, electric cars in traffic conditions 195–6
Shenzhen Universiade Centre EV Charging station 199
signposting wider demand 136
Small Business Research Initiative (SBRI) 42
social problems, identification 70–71
social procedures
environmental protection, energy conservation 20
social purposes
development of environmentally friendly products 152
development of technologies for disadvantaged groups 152

improvement of space technologies 152
linking procurement and innovation 152
social purposes in US
environmental protection, energy conservation 148
societal risk mitigation, awareness-raising and training 104
societal risks 93, 100
socio-economic goals 69
software development projects, success increase 94
software providers, small-sized efficient knowledge absorbers 227
software system (platform), centralized
for e-government services 209
sounding FCP actions 135
customers' uncertainty decrease 135
space technologies 174
'special interests', danger of corruption 13
spillover effects from innovations
market expansion 49
by public buyers 47–8
spring storm season 1985, Cimarron Doppler radar 173
stainless steel, super duplex 282
Standards of Monitoring System on NECs, Shenzhen 200
steel companies 277
storm dynamics study, Hydro-Radar 173
subcontractors: learning experience in P-51 project 227, 287
sub-sea development ROV technology 170–71
supplier relationships for innovation 79–80
supplier risks in PPI 113
suppliers, potential, procurers' unmet needs 72
supply chain interaction Rotherham case 137
supply chain management 115–16
supply chain response
Rotherham NHS Foundation Trust (UK) 128
supply chains, globe-spanning, currency risk and political risk 91
supply risk of products 239
support services for innovating firms 4
supranational rules 266
surface mining, Department of Interior (DoI) 157
sustainability criteria in tenders, not in healthcare 54
sustainable construction
e-health, protective textiles, bio-based products, recycling, renewable energy 107
Sustainable Procurement Task Force, 2005 241, 257
and link with PPOI 238–40
sustainable products and services 52
sustainable services, provision of use knowledge and technology knowledge 239

taxpayer funds from public sector 147
TCE *see* transaction cost economics
technical development, rapid digital services 71
technical expertise of purchasing authorities
for functional requirements 76–7
technical intelligence lack, policy need 60
technical judgement for assessment 16
technical skills for functional requirement development 77–9
technical support from government 77
technical testing of products, insufficient alone 77
technological backwardness 268
technological capabilities matrix by Lall 277
technological capabilities of CENPES 275–7, 286, 289–91
platform design 289
technological effort, low level, in BOSI 286

technological knowledge transference 227–8
technological learning and PPI in developing countries 268–73
technological learning in Brazil
by EPC contractors working on P-51 280–83
by Subcontractors Working on P-51 283–90
technological learning in developing countries 279
technological learning in platform projects 278
technological risks 80 100–101
non-completion, under-performance of good or service 92–3
of procured service or product 79
under- or false performance 229
technological risks and costs 101
Technological Standards of EV Charging System, Shenzhen 200
technologies, socially undesirable, underinvestment 268
technology lag in developing countries 292
technology licensing agreements 286
technology of Company A, Jinan, maturation 193
technology procurement, advanced 78
telecommunications
linking modeling and ecological information centers 169
telecommunications providers 213
tender, negotiated 52
common in works and office equipment 53
tendering process, expert advice
for inexperienced procurers 14–16, 105
tenders, assessment of 16
tenders, award criteria, innovation requirements 51
tenders, restricted 53
testing and experimental environments
piloting of new products 78
testing and piloting activities
user linking for performance verification 78–9
testing ground for innovative products 148
third-party technology 288
TIGER system 162–3
censuses of United States, Puerto Rico, Island areas 163
local and tribal governments use 163
time to market and risks 102–3
toy industry, fad-driven products, demand for 91
training programmes 82, 104
transaction bounded investments 122, 135, 138
inputs for single transaction 123
transaction cost economics (TCE) 122
contract between principal and agent 122
economics of contracting 122
transactions, characterization, uncertainty 122
translation of protocols, formats and languages 213
turbulence risks 93

UK Department for Business, Innovation and Skills (BIS), 2009 116
UK National Health Service (NHS) pilot project 124
uncertainty, absence of information 122
uncertainty reduction, of customer and supplier 137
underperformance of procured service 79
Understanding Public Procurement of Innovation (UNDERPINN) 40
survey 42–4
unemployment, fighting, US 158
US Agency for International Development AIDAIR 150–51
user-production interaction 72
users, engagement of, for functional requirements 76–7
user-supplier interactions, BYD, Shenzhen 196

vehicles, China as fastest developer of automobile industry 182
vendor qualifying condition, FAR policy (US) 151
venture capital 104
vertical alignment
  policy-makers, general management, operational level 81
vertical coordination, ministries, agencies, local government 81

waste, confidential, limited traceability
  HM Revenue and Customs (HMRC) 249
waste management and paper procurement
  traditional approach 244–9
waste paper as valuable commodity 249–50
weather radar systems, advanced, definition 172
Weather Surveillance Radar-88D 172
web-based telecommunication technologies
  (information and communication technology) 211
web-service technologies 221
wider market demand, demonstration of 136
wind speed determination 173
World Trade Organization 180
World War II, hostile aircraft and missiles 172
WSR-88D
  dual-polarization technology upgrade 173
Wuzhoulong Motors, Shengzhen
  energy-saving coaches, 2000 195
  models of NEVs 198–9
  two-layered hybrid bus model 199

Zero Waste Mattress service
  forward commitment procurement (FCP) 116

Printed and bound by CPI Group (UK) Ltd, Croydon, CR0 4YY

27/10/2024

14580411-0003